T0291291

Field Guide to Compelling Analytics

Field Guide to Compelling Analytics is written for Analytics Professionals (APs) who want to increase their probability of success in implementing analytical solutions. In the past, soft skills such as presentation and persuasive writing techniques have been the extent of teaching junior APs how to effectively communicate the value of analytical products. However, there are other aspects to success such as trust and experience that may play a more important role in convincing fellow APs, clients, advisors, and leadership groups that their analytic solutions will work.

This book introduces the formula "Analytics + Trust + Communication + Experience > Convince Them" to illustrate an AP's ability to convince a stakeholder. The "Convince Me" stakeholders might be an analytics team member, team lead, decision-maker, or senior leader that are either internal or external to the AP's organization. Whoever they are, this formula represents a concise, digestible, and above all practical means to increase the likelihood that you will be able to persuade them of the value of your analytical product.

Features
- Includes insight questions to support class discussion
- Written in broadly non-mathematical terms, designed to be accessible to any level of student or practicing AP to read, understand, and implement the concepts
- Each section introduces the ideas through real-life case studies

Chapman and Hall/CRC Focus Case Studies in Analytics and OR

This series aims to provide concise and accessible introductions to mathematical topics in Analytics and OR. Books in the series are designed to offer a primer on a topic rooted in an illustrative case study or example, and an indication of how the ideas discussed might be usefully applied by the reader. These books are intended to appeal to a mixture of professionals, practitioners and students, and will require only a relatively basic level of mathematical background.

The series is open to proposals on a wide range of topics and applications associated with operations research and analytics, including mathematical analysis, optimization, mathematical modelling, linear and non-linear programming, machine learning, and more.

Series Editors

Natalie M. Scala
Townson University, USA

James P. Howard, II
Johns Hopkins Applied Physics Laboratory, USA

Field Guide to Compelling Analytics
Walter DeGrange, Lucia Darrow

Field Guide to Compelling Analytics

Walter DeGrange
Lucia Darrow

CRC Press
Taylor & Francis Group
Boca Raton London New York

CRC Press is an imprint of the
Taylor & Francis Group, an **informa** business

A CHAPMAN & HALL BOOK

First edition published 2023

by CRC Press
6000 Broken Sound Parkway NW, Suite 300, Boca Raton, FL 33487-2742

and by CRC Press
4 Park Square, Milton Park, Abingdon, Oxon, OX14 4RN

CRC Press is an imprint of Taylor & Francis Group, LLC

Library of Congress Cataloging-in-Publication Data

Names: DeGrange, Walt, author. | Darrow, Lucia, author.
Title: Field guide to compelling analytics / Walt DeGrange, Lucia Darrow.
Description: First edition. | Boca Raton : C&H/CRC Press, 2023.
| Series: Chapman and Hall/CRC focus case studies in analytics and
OR | Includes bibliographical references and index.
Identifiers: LCCN 2022019489 (print) | LCCN 2022019490 (ebook) | ISBN
9781032065250 (hardback) | ISBN 9781032068596 (paperback) |
ISBN 9781003204190 (ebook)
Subjects: LCSH: Operations research. | Management science. | Business--Data
processing. | Information visualization. | Communication in management.
Classification: LCC T57.6 .D43 2023 (print) | LCC T57.6 (ebook) |
DDC 658.4/034--dc23/eng/20220708
LC record available at https://lccn.loc.gov/2022019489
LC ebook record available at https://lccn.loc.gov/2022019490

ISBN: 978-1-032-06525-0 (hbk)
ISBN: 978-1-032-06859-6 (pbk)
ISBN: 978-1-003-20419-0 (ebk)

DOI: 10.1201/ 9781003204190

Typeset in LM Roman
by KnowledgeWorks Global Ltd.

Publisher's note: This book has been prepared from camera-ready copy provided by the authors.

Contents

Author Biographies

Walter DeGrange is the Director of Analytics Capabilities for CANA. He has extensive experience in implementing analytical models in both the Department of Defense and commercial areas. Prior to CANA, Walter served 21 years in the US Navy as a Supply Corps Officer. He was the Director of Operations Research at several military commands as well as a Military Assistant Professor on faculty at the Naval Postgraduate School in the Operations Research Department.

Walter is also very active in analytics education. He is an adjunct faculty member at the University of Arkansas with both the Master of Science Operations Management and Engineering Management programs. He is an MBA Executive Advisor at the NC State University Poole School of Management. Walter serves as the Military Operations Research Society (MORS) Course Director for the Critical Skills for Analytics Professionals Certificate Program and he teaches the Analytics Capability Evaluation (ACE) Coaching Course for INFORMS.

Lucia Darrow is a data visualization expert with experience implementing analytical models in a variety of environments, including healthcare, manufacturing, defense, and finance. An industrial engineer by training, Lucia also has extensive experience with Lean manufacturing and the modeling of complex systems. In a professional role as a learning and development lead and as a community events organizer, she has facilitated many successful networking events, tutorials, and classes in a wide range of analytics topics. Lucia is actively involved in the analytics community through RLadies and recently co-organized the first Vancouver, BC Datajam.

She holds a Master of Science in Industrial Engineering from Oregon State University and a Bachelor of Science in Mathematics from Dickinson College. Lucia is currently Content Marketing Data Analyst with RepRisk AG, an environmental, social, and corporate governance (ESG) data science company based in Zurich, Switzerland.

Acknowledgments

We wish to personally thank the following people for their contribution in terms of inspiration, shared knowledge, and other help in creating this book:

All the analytics professionals with whom we have collaborated and learned over the years.

Natalie Scala and James Howard for allowing us to write this book in their analytics series.

Norm Reitter and Jennifer Ferrat for their help in selecting the perfect title.

Al Roth for his passion for the Kidney Exchange program that provided many of the stories to demonstrate the concepts in the book.

CHAPTER 1

Introduction

A NALYTICS HAS influenced the world in many ways over the past half century. With advancements in computers, software, and algorithms, analytics will continue to grow and transform the world of the future in ways we can only dream of today. The impact has been felt in many areas including business, medicine, sports, defense, transportation, and education. With the increased use of analytics, it is important to understand how to design, develop, communicate, and represent analytical models and solutions. While the application of analytics grows, the fact remains that the mathematical principles that analytics is based on, are only understood by a small percentage of the population known as Analytics Professionals (APs). The APs must be able to convince the larger population that these algorithms work and guide their application.

1.1 DEFINING AN ANALYTICS PROFESSIONAL

An AP is someone who uses data and math to improve decisions. In this book, we are including all of the following titles into the broader description of AP: Data Scientist, Operations Research Analyst, Data Engineer, Data Architect, Data Storyteller, Machine Learning (ML) Engineer, and Artificial Intelligence (AI) Engineer. This definition encapsulates any field that uses data and advanced math to yield insight. We do not refer to any specific job title due to the fact that these roles change over time and that sometimes the meaning of the titles vary from one organization to another. New titles are continuously created and the connection between all of these roles and what they entail evolve over time.

Could many of these lessons and techniques that we cover in this book be used by a spreadsheet analyst? Absolutely. However, we will focus on use cases where the mathematical techniques are more complicated in nature, such as mathematical optimization or data science.

1.2 DESIGN OF THE BOOK

This book is designed for the individual AP. Most other books dealing with analytics tend to be focused either on the analytics team or the organization that is using analytics. When trying to apply the insights provided by these books at either the team or organizational levels, the lessons are difficult to scale down to a single person. For example, the requirement for senior leadership at an organization to support the implementation of a major analytics project is a common recommendation. Who should be in charge of accomplishing this and how can an individual AP make a difference? Another recommendation that is commonly highlighted is an increase in the level of communication. When implementing this insight for an entire team, then what team member(s) need to sharpen their communication skills?

This book focuses on the contribution of the individual AP to the entire project. In analytics, as with the rest of life, a person can only control their own actions. Focusing on the individual AP is a unique perspective. While steps taken by one person can increase the probability of success for any given analytics project, these actions alone cannot guarantee success. As such, we recommend pairing the lessons from this book with approaches from the team and organizational level to help shape an analytics culture.

The book is organized using the equation in Figure 1.1 to illustrate the individual AP's ability to convince another person or group of people.

Analysis + Trust + Communication + Experience
>
Convince Them

Figure 1.1 Convince them equation.

The other person or group may range from fellow analytics team members to leadership within their own organization to world leaders. Using the equation allows us to emphasize the point that the variables on the left-hand side are different for every audience. No two people or group of people have the exact same experience, education, and relationship with the AP. Likewise, no two APs have identical backgrounds and strengths. Therefore, factors on both sides of the equation are changing over time and vary from individual to individual.

1.3 PURPOSE OF THE BOOK

Compelling analytics answers the question "so what?". It presents focused, engaging results that show the impact potential of an analysis, while maintaining truthfulness and transparency—without skipping over or distorting limitations to the analysis. This integrated approach to analytics not only brings stakeholders into the driver's seat of an analysis, but also makes them invested in where it can lead.

The focus of this book will be largely on analytics in a business context, which aims to help companies understand their operations and perform better. However, many insights on analytics implementation and the case studies woven throughout the text are applicable in other contexts such as academic research.

Any complex process is built by a series of steps. These steps are dependent on each other for overall success. Analytics is a process that takes data as an input, uses math modeling as a tool to transform the data, and outputs information that can be used in other formats such as data visualization or decision support. When you break any process down to component actions, there is an inherent risk of over-emphasis on the success of one of the actions. In the case of analytics, these actions include such things as communication and data visualization.

An example of this concept is when a new driver gets behind the wheel of a car for the first time. They are bombarded with new mechanics such as checking mirrors and rules like being able to make a right hand turn on a red light. Although a very important step, no one just practices turning a wheel left and right by itself and there is always a context on how and when to turn the wheel. For instance, the style of rotating the wheel slowly and deliberately for parking differs from the method of veering quickly to avoid an accident. Although each step is important, no process should be overemphasized or considered in a vacuum.

Even an optimized slide deck with excellent data visualizations will not effectively communicate the analytics in every situation and for every audience. Knowing when and how to change these tools to fit the context is the magic of the profession. APs must master this skill over time and through many attempts—both successful and not so successful.

Analytics is a tool much like a hammer. The hammer can be used to build a beautiful house and it can also be used as a weapon. This book teaches tools and techniques to enhance the use of analytics to induce change. We hope this change is positive and enhances the lives of individuals, organizations, and the world as a whole.

1.4   KIDNEY PAIRED DONATION CASE STUDY

As of March 2022, there are approximately 90,000 individuals [2] on the waiting list for kidney transplant in the United States (Organ Procurement and Transplantation Network). In comparison to most other organ transplant procedures, kidney donation is unusual in that live donations are possible as humans can live a normal life with just one kidney. The long-term benefits of kidney transplantation are well known. A successful living donor kidney transplant increases the lifespan of both the recipient and donor, but also brings about other benefits in the long term, including a reduced risk of heart disease.

Since the development of medical technology to allow for kidney transplants in the 1960s, donations by living donors were performed in the same location and included only the donor and the recipient. Kidney exchange or paired donation connects incompatible pairs of recipients and donors to increase the living donor pool. In the simplest form of exchange, an incompatible donor-recipient pair (for instance a father and daughter with differing blood types) are matched with another incompatible donor-recipient pair. The donors then give their kidneys to the other recipient in the match. In order to ensure involved parties respect the agreement to exchange organs, the transplants must occur simultaneously. As this donation exchange or chain scales up to incorporate several recipients, the complexity multiplies.

The exchange process is sometimes supported by Good Samaritan donors, who provide an organ without connection to an active participant on the waiting list. Even though by the early 2000s advances in medicine could allow for other combinations of donors and recipients, there were logistical, legal, and ethical obstacles that needed to be addressed before this process was possible. These challenges provided an opportunity for a team of APs to propose a solution using a mathematical approach adapted from economic theory. Their work has changed countless lives and made the unthinkable possible: the longest donation chain to date incorporated 70 surgeries and 35 transplants [2]. Throughout the book, we explore the analytics, trust, experience, and communication dimensions of this complex problem to understand how the team of researchers were able to use analytics to enable change.

Analysis

A NALYTICS TOUCHES many disciplines, from math and statistics to operations research and industrial engineering. The use of analytics is found almost everywhere we look: in academic research, tech, finance, logistics, and healthcare. If you asked a digital marketer, an operations research analyst, and a data scientist to define analytics, you will probably receive three relatively different answers on how analytics is applied; however, the common thread between the three answers would be that analytics uses math to turn data into insights.

The Institute for Operations Research and the Management Sciences (INFORMS) defines analytics as the scientific process of transforming data into insight for making better decisions [11]. Analytics encompasses not only the final model, but also several contributing processes including collecting and exploring data, defining scenarios, setting model features and constraints, tuning, and presentation. The power of analytics is then bringing actionable insights out of large amounts of data that are difficult or impossible for humans to comprehend. To this end, there's always a balance to be struck between the usefulness of an actionable insight and understanding the data that generated it.

DOI: 10.1201/9781003204190-2

An endless number of resources on the subject of analytics methodology exist which we will not detail in this book. In this chapter, we instead review the analytics workflow from start to finish, focusing on the practices that make or break the long-term use of an analytics solution. We look at several examples of analytics in practice and present ways to make analytics solutions more interpretable, robust, and in alignment with existing infrastructure.

2.1 REAL-WORLD IMPACT OF ANALYTICS

Despite the recent surge in interest in analytics, there are a few features that make some business problems unfit for analytical solutions. Problems that are not well suited for analytics are problems that are hard to explain or constrain with explicit rules. First, if you cannot clearly describe what the problem is to a coworker, it will likely be difficult to translate the scenario to a computer. Second, lacking supporting data can make analysis challenging, if not impossible. However, both of these challenges can often be overcome by time and cultural support of analytics. If data does not yet exist, a process can be set up to collect it. If the business problem is too broad, it can be narrowed by workshops with stakeholders or by exploration of historical data. Finally, the third and most difficult challenge for analytics can be solving problems that require great creativity. With current tools, creativity in the human sense can be difficult to simulate and some problems are best solved with partial or full human oversight.

In the following section, we'll review a few examples of analytics in practice.

Big data at Netflix

The popular video streaming service Netflix is often regarded as a leader in recommender systems. By suggesting enjoyable content, Netflix increases user retention. How does Netflix determine the right content to show and ensure the right amount of variety in its recommendations? When Netflix was primarily a video rental service, the videos a customer checked out and how they rated them were the crucial data points driving their algorithms. As their platform transitioned to streaming, Netflix gained access to more granular data: what users watch, when they watch it, for how long, and on what device. The wealth of information about each user's preferences feeds into a variety of algorithms that determine what is shown on the segmented rows of the browse page. The trending row incorporates spatiotemporal data such as movies about upcoming holidays or titles that are popular in the user's location. Randomized, controlled experiments called A/B tests are core to how Netflix understands and influences user retention. These tests can provide statistical evidence for questions such as "which layout is best?" or "which title should be shown to a user?"

To solve a piece of this puzzle, Netflix even hosted a competition from 2006 to 2009, called the "Netflix Prize," for the best collaborative filtering algorithm which predicted how a user would rate content [18]. The winners of the open algorithmic contest beat Netflix's baseline algorithm by a little over 10 percent. While Netflix was able to make improvements based on contestant contributions, the winning ensemble method turned out to be too complex to scale and productionize.

Analytics shapes the world around us

Data has long been a determining factor in urban planning: influencing decisions such as where public transit should run, how to ensure equitable access to nutritious foods, and how urban space should be allocated. While supporting data for these kinds of decisions often came from observational studies and surveys, big data is changing how cities are developed. Access to sensor data and rich spatial data allows planners to create a digital twin of a city. The digital version is a complex simulation environment to test planning decisions.

In 2020, the city of Pittsburgh, Pennsylvania, released a data-driven pedestrian safety plan [4]. Analysis of historical data on crashes involving pedestrians allowed the city to understand the most important factors influencing the likelihood of an accident. The development of the pedestrian network, primarily where new sidewalks should be built to maximize public safety, is now guided by analysis of spatial data.

Data for the win in racing

Formula One racing is one of the most exciting competitions in the world as shown by the race car in Figure 2.1. It has been around for over 100 years and continues to grow annually. As the sport has matured, teams have honed their use of analytics in order to stay on top of their game.

Figure 2.1 F1 car racing on track. Davis, 2019.

As of the 2021 F1 Grand Prix season, the Mercedes F1 team's engineers, analysts, and race strategists analyze data collected from more than 200 physical sensors in its cars. Additional data is continuously gathered from the factory and track—an average of 20 TB/day.

One example of analytics that racing teams use to help their driver win is prediction models. These models are used to predict racing strategy decisions, fuel load, and tire wear. The data is then transferred to other parts of the racing team to aid in decision-making. One form of data analysis racing teams put into practice is "correlational analysis," through which algorithms determine the relationship between two variables. Correlational analysis can also be done with categorical data that racing teams will use to compare different strategies and see which one performed better.

Another interesting area of racing analytics is the use of simulation models to improve upon track performance as well as car design. Simulation models are created to test changes in car design and racing strategy to see if they would alter performance. The models are an extremely important part of racing analytics because they can help racing teams save time and money as well as ensure the safety of their drivers through better decision-making around race day strategies.

Analytics also influences how racing teams use data to design more efficient racing cars. Racing teams use simulation models to test the potential of their designs and see how they compare with the designs of other teams. With simulation data, teams can determine if changes in car design will increase performance or not, which is an important factor for racing teams when designing new cars.

Racing analytics is an important tool racing teams use to stay one step ahead of their competition. From correlational analysis, data modeling, and simulation models for both track performance as well as car design, racing teams rely on these tools to make better decisions in the racing industry.

2.2 TYPES OF ANALYTICS

While analytics can take many forms, there are three generally accepted classifications: descriptive, predictive, and prescriptive.

Descriptive analytics provides insight into what has occurred in the past, using summary statistics and data mining to describe historical data and trends. Examples include creating a dashboard to describe the last month of performance for a product or reporting on engagement on a social media post. Descriptive methods help uncover relationships in the data and are often an important first step before more in-depth analysis.

Alternatively, predictive analytics uncovers patterns in data, using statistical modeling techniques to assign probabilities to future events. Predictive methods have become integrated into our daily lives and many examples can probably be found on your smartphone today. For instance, the models behind applications that determine the species of plant from a user-uploaded photograph or identifying the song playing on the radio uses this approach. Predictive solutions often include a measure of certainty or likelihood that the given prediction is accurate.

Lastly, prescriptive analytics guides decision-making, using mathematical models, optimization, and simulation to explore courses of action. In these methods the focus is on finding an optimal or near optimal solution. A sample use case of prescriptive analytics might be a delivery service determining the optimal route to minimize fuel use or delivery time for their drivers.

TABLE 2.1   INFORMS Body of Knowledge Domains of Analysis

Domain	Phase
I	Business Problem (Question) Framing
II	Analytics Problem (Question) Framing
III	Data
IV	Methodology (Approach) Selection
V	Model Building
VI	Deployment
VII	Model Life Cycle Management

2.3   THE ANALYTICS PROCESS

In practice, analytics projects rarely run in a linear fashion, but rather steps may need to be revisited as conditions change and the project progresses. Every analytics endeavor should begin with a clear objective, even if that objective is simply to perform exploratory data analysis (EDA). Next, the supporting data must be identified, formatted, and prepared for analysis. After analysis, results, and insights can be drawn from the data. As a project progresses, there may be a need to reassess objectives or return to the data collection stage as the problem may change over time. The analytics lifecycle provides a roadmap for the development of analytical solutions. The analytics process section walks through the steps of the analytics lifecycle, focusing on actions an individual AP can take along the way to ensure project success.

We structure our review of the analysis process around seven domains identified by the INFORMS Body of Knowledge presented in Table 2.1 [11].

2.3.1   Framing the Problem

While analytics projects sometimes spur from discoveries in data exploration, most frequently projects begin with a business problem that requires an analytical lens. Thus, at the onset of an analytics project, an AP often needs to translate a business problem into an analytics problem. A business problem may be multifaceted, open-ended and ambiguous. It is the job of the AP to distill it into one or more analytics problems that are clear, targeted, and measurable.

Business problem

While it is natural for APs to focus on how to solve a problem, in the early stages of a project it is important to understand the "why" behind an analysis. The "why" comes from many sources: the stakeholders, the process, and the data. Communicating with several stakeholders, from subject matter experts to data experts, provides a more complete picture of the problem at hand. Different stakeholders may have different requirements and it is the responsibility of the AP to clearly communicate which requirements will be addressed.

Stakeholder communication is only one piece of understanding the problem. A seasoned AP may also consider inexplicit features in the problem through their own research and observations. A key principle in the Japanese theory of lean production is the idea of "go to gemba," which means go to the place where the work occurs. In process improvement, this entails respectfully observing work and learning from the individuals who are closest to the process. While your analytics project may not always involve a physical process, this method is incredibly valuable for APs. By observing a process directly and communicating with people you may not traditionally have access to during an analytics project, you may uncover insights or explanations that are not readily apparent with data alone. For example, for a digital marketing analyst investigating the impact of a particular campaign, this could involve supplementing click data with customer surveys that highlight nuances in the project.

Analytics problem

As with any project, analytics projects should be benchmarked to clearly show their impact on the problem at hand. As you plan your analytics approach, you should also define project goals and what "success" looks like. Accordingly, the analytics results should directly translate into a business result.

Translation of the analytics problem is not always easy. The business problem may be framed in terms of symptoms rather than root causes. APs need to drill down to find the factors that are driving the business problem so they can come up with a hypotheses to use when framing the analytics problem. A good AP can take a complex business problem and turn it into a series of well-defined analytics problems that a mathematical model can tackle.

2.3.2 Requirements Gathering

At the requirements gathering stage, the AP meets with the clients and stakeholders for the analytics project. Clients could be internal, as another team in your organization using your product, or external, as with a consulting project. Requirements gathering can occur in an informal meeting, a workshop, or through surveys and is key to establishing a baseline understanding of the problem and potential analytics solutions. It is also an opportunity for the AP to manage client expectations by clearly stating what is possible given the parameters that are outlined in the session.

Before meeting with the project stakeholders, define the elements you need for the model. Generally, these would include the objective of the model, the levers that someone could change to test different scenarios, and the assumptions that will inform model development.

These elements will differ from model to model. For example:

- For statistical analysis, this may include: level of significance required.

- For a simulation, this could include: an initialization routine, assumed distributions, and an ending condition.

- For an optimization, this could be: the objective function(s), decision variables, and constraints.

An often overlooked aspect of requirements gathering is planning for eventual deployment of the model. If this is an analysis that will be used regularly, it would likely be beneficial for the model to be integrated into an existing technical infrastructure with additional models, tools, or applications. It is highly recommended to consult with Information Technology (IT) and software development professionals within your organization to understand the existing infrastructure and how analytics projects have been deployed in the past. It is easier to consider the plan for deployment from the beginning than to patch together a deployment at the end of the project. Questions to ask early include:

- How will supporting data for the analysis be stored, accessed, and updated?

- What programming languages, packages, and frameworks appropriate for the project will integrate well into the existing technical structure?

- If your preferred analysis tool will not integrate seamlessly, what actions can you take to enable its integration, or what other tools can be considered?

- How will you update the model and receive feedback about its performance?

- What support, if any, will you need from your IT and software development teams? Keeping these groups in the loop regarding your analysis will reduce friction when it is time to deploy the model.

Requirements gathering can also be a useful time to understand the stakeholder's style, aesthetics, and preferred mode of project delivery. The following considerations may help you to better understand their preferences:

- Are there past projects that were particularly successful in the eyes of the project stakeholders? What features did they like about these projects?

- If you are working together for the first time, consider preparing examples of your past work: are there elements the stakeholder would like to see replicated in this project?

- What publications or applications do they like? Can you bring elements of these resources into the presentation of your analytics solution?

Following a meeting with stakeholders, you may need to do some refinement and prioritization of the requirements to meet the time constraints of the project. Document the final requirements and schedule for delivery and share them with the stakeholders involved in their development. As analytics and business needs are dynamic, expect that there may be some changes to the project requirements over time. Documenting requirements can be achieved using spreadsheets, online documents, or presentations. If you are using presentations make sure there is enough insight to provide context. The notes section on a slide is an excellent place to capture these critical pieces of information.

Surprise! Deviating from the plan

What happens when you uncover something in the data that deviates from the original business problem? In 2004, Linda Dillman, then Wal-Mart's chief information officer, asked a team of APs to predict how many emergency supplies their coastal stores should stock in advance of a hurricane [20]. It was expected that shoppers would stock up on items such as flashlights, bottled water, and first-aid that make up a traditional disaster preparedness kit. The APs turned to historical data to explore past trends in pre-hurricane baskets from their regional stores. To their surprise, one item stood out with an increase in sales at a magnitude of seven times its normal sales rate: strawberry Pop-Tarts. By widening the analysis scope and looking for patterns in the full product range, Walmart analysts were able to uncover unexpected trends and shift the focus of their project to increase revenue and meet customer desires. Insights from historical data are particularly powerful when they highlight phenomena that go against human intuition.

2.3.3 Working with Data

Data collection and cleaning

Whether it's historic, streaming, notional, or synthetic, all analytic models start with data. As the all too familiar phrase "garbage in, garbage out" implies, poor quality inputs will yield poor quality outputs, regardless of the strength of the model. Data quality can make or break the accuracy and longstanding impact of a model. The quality of data is not fixed and can change or degrade over time. As you identify and set up a collection process for data, engage appropriate stakeholders and subject matter experts. Those who are closest to the data will have valuable insights on data collection tools, accuracy, and common errors, as well as notable trends in the data. As you start gathering data, consider if and how you will access this data in a month's, or even a year's time.

- How will the data be stored? Does your project require a database?

- How often will the data be updated?

- How will the data be labeled? What metadata can be used to uniquely identify the data?

- Who will need access to the data in the long run and how can you enable it?

As you establish data collection methods, the risk of bias should be mitigated as it can easily be introduced into the model through the data collection process. With this in mind, the data gathered should be representative of the situation you are trying to model. Early attempts at facial recognition software provide a clear example of what happens when non-representative data is used for training. Determined by training sets largely composed of male, white faces, the models had a much higher error rate for women and people of color. In 2018, the Gender Shades project highlighted discrepancies in accuracy between 20 and 35 percent for leading, and largely accepted, facial recognition technologies [10]. The study showed the "coded gaze" (algorithmic bias) of those developing and deploying AI technology, while highlighting the risk of overlooking bias in data used for training.

2.3.4 Data Formatting and Storage

Throughout the data collection and formatting stages, designing for traceability reduces complexity and clarifies key questions for future validation. It is easy to overlook documenting data source and transformation information at the start of a new analysis, but this information is key for reproducibility.

The analytics data pipeline consists of all of the transformations that take data from its original source to the output at the end of the modeling process. A defined data format and method of storage enables ease of use by not only APs, but also by individuals in your organization that may need to access the data. Putting effort into the format and accessibility of data takes some time up front, but will reduce the need for rework in the future. Consider the data type (e.g., structured or unstructured) as well as the frequency at which the data is updated (e.g., monthly, daily, streaming real time) as you establish a data storage method. Meta-data can be used to create a concise description of the data for easier organization and access within an organization.

2.4 SELECTING AN ANALYTICAL APPROACH

As you weigh different analytical approaches, consider the following factors presented in Table 2.2.

APs often have their own experiential biases toward analytics methods: this could be favoring what you've used frequently in the past or eagerness to use a new method that is popular. With the

TABLE 2.2 Analytical Approach Factors

Factor	Question
Feasible	Is the business problem amenable to an analytics solution?
Informed	What has been done to solve this problem in the past?
Aligned	Does the solution selected solve the analytics problem at the core of the business problem?
Ready	Is there data available or could data be collected to support an analytics solution?

advent of methods such as deep learning, the latter is becoming increasingly popular. Using AI to answer a simple business question is like using a sledgehammer to crack a walnut. Always consider whether the business problem at hand necessitates the selected analytics method and validate with other APs if possible. Depending on the type of analytics model you are building, it may be advantageous to take several types of models into the next stage to test and compare their performance with real data.

2.5 MODEL BUILDING

2.5.1 Defining Model Components

Business rules help a mathematical model approximate behavior in the real world. While specific model components will vary based on the type of analytics you are performing, these components can fit into three general categories:

- Goal: What are you trying to summarize, predict, or optimize?

- Variables: What elements of the problem can change? Which elements' behaviors are project stakeholders most interested in?

- Constraints: What limits or rules exist that will impact the model?

2.5.2 Efficiency: Don't Repeat Yourself, Don't Reinvent the Wheel

DRY (Don't Repeat Yourself) is a concept from software development that aims to eliminate rework. The benefits of the framework are two-fold: not only is the burden of development time reduced, but also the risk of overlooked errors in the code is managed. For example, say you have written code to calculate a key metric. If you need this same code in another section of your model, it would be unnecessary to re-generate the function from scratch. Often the easiest way to utilize the same functionality again is to copy and paste the code into the new location creating multiple instances of the same code.

Now say a few months later, a colleague identifies an error in your code and asks to change the metric. Could you guarantee that you would remember to update every instance of the erroneous code? Under the DRY principle, you would have created a single function, referenced in several locations. With this method, when you make changes in the code, it is only required in one location. DRY is superior by saving time and ensuring that the code works the same throughout the code base with only one update.

Developer time is a crucial resource on analytics projects. After you sketch out an approach, do some research to understand if and how this problem has been tackled in the past. Consider if a similar problem has been solved before in your department, organization, or by others in the analytics community. An existing analytics solution may be able to solve part of your problem you may be able to learn from others who have attempted to solve a similar problem in the past. For instance, if you were performing text mining, you likely wouldn't code your own function to stem words, but rather utilize a solution from a well-regarded software package. Mature solutions that have already been developed by your organization or others in the community have the benefit of being soundly tested and optimized for performance. By utilizing existing solutions for common parts of your problem, you free up time to focus your attention on specializing your approach to the aspects of your problem that make it unique.

2.5.3 Speed vs Accuracy: When Simple Is Better

Accuracy is not the most important outcome for all stakeholders. For example, sometimes full explainability of the approach is non-negotiable. In the requirements gathering stage, a timeline for delivery of the model is established, often accompanied by a threshold for acceptable model runtimes. These two constraints dictate how detailed, how accurate, and how optimized your model should be. It is easy to assume that more time spent and more details modeled will always equate to a more precise and powerful solution. Naturally, one would expect the more information a model has regarding the real world, the more accurately it would be able to model reality. However, there are several factors to consider when determining how detailed your model should be.

Overfitting is a common issue in predictive analytics, in which the model is too closely fit to the training data. While the model may achieve high accuracy on the training set, it may not perform as well with the test set or with new data. The bias-variance trade-off refers to the relationship between two sources of error that impact the generalization of a model. While bias error comes from the inability of a method to capture the true relationship in data, variance error comes from oversensitivity to fluctuations in the data. An overfit model has high variance and low bias. On the other hand, underfitting occurs when a model has low variance and high bias. Balancing these two factors is key to producing accurate predictions across datasets.

2.5.4 Model Design Trade-offs

A robust model is not rendered ineffective by changes to the data or modeling scenario that are expected. For instance, data drift, the phenomenon of the distribution of a supporting dataset changing over time with natural shifts in a process or market, is a known risk for predictive models. Robust modeling approaches would monitor this risk and have a procedure in place to update the model in response. Sensitivity analysis captures uncertainty or real-world modeling parameters such as demand, temperature, or prices. It can determine how robust an optimal solution is to changes in parameters and should be performed for most prescriptive models. Even if the models do not experience changes to the extremes under which they are tested during this process, sensitivity analysis increases the AP's familiarity with the behavior of the model.

Representative models start with representative data. If a model is not trained on or developed with data that is representative of what it may encounter in production, it will not perform well on real data. A representative sample should reflect the characteristics and variability of the real-world dataset. Thus the data must be large enough to accurately capture intricacies within the distribution. In some cases stratified sampling may have advantages over random sampling due to the ability to ensure the inclusion of subgroups. In predictive analytics, cross validation is a valuable tool to improve model performance. Cross validation uses resampling to extend the usefulness of a dataset.

Model flexibility is critical for long-term success. As business needs change, an analytics model will be much more long lasting if it can be adapted to some degree of changing requirements. Model flexibility can be achieved by designing for extensibility, as in designing a model to accommodate future growth and requirements. An unexpected way to increase model flexibility can often be found in simplifying a solution. Solutions that model a process to a tee, incorporating minute details, falter when presented with new processes or changes to the initial process. Simpler models are generalizable, and in turn more flexible to implement. For instance, in prediction, reducing dimensionality of a problem allows for focus on the most impactful features. Simpler solutions may have minor drawbacks when viewed from the narrow lens of the current problem, but will pay dividends on future applications.

2.5.5 Testing

After a model has been constructed, it should be tested to ensure it functions correctly and provides the accuracy of results required by the client. The testing process is often referred to as verification and validation. Model verification answers the question "does the model perform as expected based on how I programmed it?" while model validation answers the question "does the model accurately represent the real world?" Verifying a model ensures that it has been implemented correctly. The process could include code reviews, calculation reviews, and testing model behavior under different or boundary scenarios. Validating a model can be trickier, especially when modeling future events or events that have never occurred. Validating a model should incorporate reviews with subject matter experts, either through direct analysis of the results or through comparison to similar model results in the field. In order to stand the test of time, analytics solutions should not only behave correctly for a single scenario, but also be robust, representative, and flexible to handle future scenarios.

The process of testing can look quite different for different types of analytics models. For descriptive analysis, testing may involve code reviews, comparisons to past analysis of the data, and reviews with data experts. For predictive models, train on data that is of the same structure and degree of cleanliness as the data the model will encounter in production.

Discrete event simulation models can pose a particular challenge for validation and verification due to their complexity, interconnectedness, and layered Graphical User Interfaces (GUIs). All simulation software programs are equipped with a set of default settings and distributions that may impact your results if they are not known and taken into account. Make note of the default settings so you are not taken by surprise and can change them if needed. In the requirements gathering stage, you likely identified key behaviors in the system and potentially even assigned appropriate distributions to model them. As you test the simulation model, review settings, behaviors and results with subject matter experts. When possible, a simple input-output validation on historical data can help to gauge the accuracy of your simulation in reproducing reality. If you are modeling a scenario that cannot be observed, it is recommended to try to compare your outputs to results of other simulation models of similar systems.

2.6 DEPLOYING A MODEL

Analytical models fall into two major categories in terms of their deployment requirements. The first is models that help to understand business questions better but are not relied on for regular use. These models often retain an "analyst-in-the-loop," meaning they stay under the supervision of the analyst and are not frequently viewed by an external user. Runtime minimization in this instance isn't of high importance as the model will likely only be accessed once per month or year depending upon how often they're required to be used. Another type falls under "applications" found online where APs must pay particular attention to algorithm runtime. For these applications, speed matters most when there could potentially thousands of user requests coming through at any moment. The best example of this is seen in Google search algorithms.

When the model is deployed, it crosses the threshold from analyst-in-the-loop testing to access by the end user. At the deployment stage, business decisions begin to be made more regularly off of the model results. Depending on the user base, the level of robustness required for deployment may vary. In some cases, an analyst-developed application may be sufficient, while in others, support from software developers may be required to ensure the application is stable. The vast majority of models never become fully deployed and integrated into the business. The deployment process is incredibly time intensive and difficult to realize if productionizing a model has not been planned from the beginning of development.

Run time is a challenge that APs encounter frequently with model deployment. While a ten second run time is perfectly acceptable for a single analyst running a model, this may be too significant for a model on a server that may have multiple users accessing the product at once. Using benchmarking tools available in most programming languages, you can target and reduce expensive operations within your model. At a certain point, run time can no longer be reduced without changing the underlying structure of the model. If a fast run time is essential or the product will be accessed by many users, consider spending time to parallelize your program or utilize a faster, less optimal method of arriving at a solution, such as a heuristic.

2.6.1 Providing Analytics to the End User

Models are most useful if they are accessible by the end user. While it was once quite standard for models to require an analyst-in-the-loop in order to run an analysis, it is becoming the norm for stakeholders to have direct access to analytics. Direct access to the model allows those with subject matter expertise to be more active in the experimentation and testing processes. It also allows analytics products to speak for themselves, proving their usefulness by quality results. With this direct access then comes greater requirements for a user-friendly and aesthetically pleasing User Interface (UI), as well as a basic requirement of stability for the product. Models need to be wrapped in an environment or application that shares the right level of technical detail.

Production with respect to analytics is defined as any mathematical model that is being used regularly, with a stable code base. Calling a model in production could apply to a model that is run once a year to provide a recommendation on pricing in support of procuring transportation or a search algorithm that runs 10,000 times per minute on a real estate website identifying houses for potential buyers.

With the advent of more advanced data science production tools, the role of the AP is increasingly inching into software and web development territory. APs thrive in rapid, experimental environments. Many APs do not have formal training in software development and ideas such as unit or load testing are completely foreign. The transition from experimentation to creating a stable, reliable product for users is a significant shift in priorities.

If you work in an organization without IT support, there are many online resources to help to get your project up and running. If you do work with IT and software engineers, your organization likely has a more robust data pipeline. Throughout your analytics project, consult with these teams to ensure your product will be able to fit into the existing infrastructure. If your analytics project is available on a familiar platform and generated with the help of software development experts, it is much more likely to be used.

Analytics models in production are generally assumed to be fully automated. Fully automated means that the stream of input data, model runs, and results generation can all occur without analyst oversight or intervention. When determining whether to automate a step in your analytics project, weigh the time it would take you to automate that step against the time it would take you to repeat the action every time the model is run. If the time to automate the step is lower and it does not require complex human decision-making, then it is likely a good case for automation.

Once a model has been deployed, there is usually still some work to do to get the model integrated into the organization. While deployment integrates the model into the technical pipeline, it is not yet integrated into the business.

2.7 POST DEPLOYMENT: MODEL MAINTENANCE AND DATA MANAGEMENT

After deploying the model into production, important factors such as updating and maintaining the model and data become the focus of effort. The process of updating the model and data should be as automated and streamlined as possible in order to minimize the impact on operations. Both periodic scheduled reviews and performance triggers can be used to inform APs that the model or data require evaluation.

2.7.1 Versioning

Versioning your analysis and its supporting data is incredibly valuable for reproducibility and maintaining model code. While there are many constructs and methods, versioning refers to simply naming and storing a record of your analysis. If you are trying to communicate to someone what has changed in an analysis, versioning can provide notation of what each form of the document includes. Version control software, such as GitHub, help to simplify this process for code and models. For true reproducibility, the supporting data for an analysis should also be versioned.

The term Minimum Viable Product (MVP) refers to an initial version of a model that meets the basic requirements agreed upon by the client. A MVP is generally reviewed by stakeholders to ensure that the model is on track to meet the needs of the analysis. Sharing a MVP stage model allows you to share your progress and gather valuable feedback about what can be improved in the model and how stakeholders would like to receive the results.

In software engineering, there are three major stages:

1. Development – Version numbers 0.1–0.5

2. Alpha and Beta Testing – Version numbers 0.6–0.9

3. Release/Support – Version numbers 1.0–higher

APs can apply the same strategy for both model development and data.

Data versioning becomes extremely important when considering applications with machine learning models. The potential exists for multiple training and test sets of data to produce completely different hyperparameters. Tracking the relationships between the data version and the model version can become complex and time consuming. The challenge exists both during and after development.

2.8 WRAPPING UP ANALYTICS

As we can see from the preceding sections in this chapter, figuring out the analytical solution is a daunting series of tasks. The AP must practice this process repeatedly to learn about the large decisions that make or break an analytics project and also the small details that can save precious development time in key areas. Although you can't have an analytics project without analytics, this is only the first of our five elements. Let us take a look at how analytics was applied in the Kidney Exchange model example.

2.9 ANALYTICS EXAMPLE: KIDNEY EXCHANGE MODEL

For many years the matching of kidney donor-recipient pairs was achieved on a one-for-one basis. The one-for-one matching limited the number of donations nationally since there had to be an exact match. In many cases eager donors are not a match for a loved one, so the recipient had to wait for a match using the national donor list. Incompatibility could sometimes be overcome by exchanging a kidney with another incompatible patient-donor pair. In this scenario, the operations would occur simultaneously to ensure neither of the donors would renege from the exchange. A comparison of both processes is provided in Figure 2.2.

A team of researchers began to question why the exchanges must occur simultaneously and explore ways to improve donor-

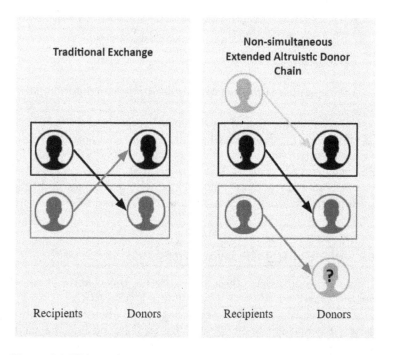

Figure 2.2 Kidney donor pairs vs the chains.

recipient matching. They used an integer programming algorithm to identify optimal cycle and chain exchanges, with the goal of maximizing the number of transplants that occur. Together with the Alliance for Paired Donation (APD), they pioneered Non-simultaneous Extended Altruistic Donor (NEAD) chains, a new donation method that increased the number of transplants possible by using chains instead of cycles. [7]

In the NEAD method, an altruistic donor kicks off the chain by donating a kidney. In turn, the recipient's donor then gives their kidney to another in need, and the chain continues. Thus, each pair gets a kidney before they give one. Introducing an altruistic donor allows the simultaneity constraint to be relaxed, enabling much longer sets of donations. The use of operations research also helped to find matches in these chains for highly sensitized patients and difficult to match blood types, who typically would face much longer wait times.

The new proposed method was initially met with skepticism from the medical community. Because there is no guarantee of a donor holding up their end of the deal, experts wondered if NEAD chains would be effective in practice. Beyond this concern, a 2009 study argued that the implementation of NEAD chains would not yield more transplants. The researchers need to overcome this skepticism to implement their model.

The research team responded to this study, pointing out that one key constraint had been included: chains were set to always contain under three pairs. While this was a common constraint for traditional exchanges, the unique advantage that NEAD offered was that it would enable much longer chains, impacting more people. The research team used a simulation to prove the impact of the chains with the chain length constraint removed. They found that NEAD chains would not only yield more transplants, but also reach more traditionally hard-to-match patients.

Such kidney exchanges have since been accepted by the medical community and emerged as a standard mode of kidney transplantation in the United States. The model used by the APD and similar transplant groups relies on a continuously updated pool of both donors and individuals in need, weighted by their urgency. In the chapters that follow, we explore more dimensions of this complex problem.

EXERCISES

2.1 How do you see the influence of analytics in your day-to-day life?

2.2 Consider a recent analytics project. How did you define the requirements with stakeholders? If you could repeat this process, what would you have done differently?

2.3 What special considerations would you take when validating a discrete event simulation model?

2.4 Map out the steps of a recent analytics project. What steps do you think could be automated through a function? Are there any aspects of the project you do not think could ever be automated? Why or why not?

2.5 What challenges might arise when deploying a model?

FURTHER READING

Cochran, edited by James J. (2019). *INFORMS analytics body of knowledge.* John Wiley and Sons, Inc.

Spiegelhalter, David. (2020). *Introducing The Art of Statistics: How to Learn from Data.* Numeracy 13, no. 1

CHAPTER 3

Trust

"I used to take on trust a man's deeds after having listened to his words. Now having listened to a man's words I go on to observe his deeds."

Confucius
The Analects

WHEN THE research team began advocating for their new method to improve kidney donation matching, the transplant culture and infrastructure only allowed for simultaneous exchange in pairs. Two operating rooms needed to be prepared at one donating medical facility: one for the donor and one for the recipient. Due to the simultaneous surgery constraint, less than mathematically optimal model results were initially used to gain trust. Using simulation results, the research team showed the impact potential of using longer chains and began to gain advocates from the medical profession. These small wins started building trust in the community to eventually pave the way for more significant changes.

Legal barriers to implementation existed based on the definition of compensation for a donated organ. A US law was eventually clarified to permit cross-donation and later chain donations to occur, however this concern remains a challenge for the global adoption of the technique. The future implementation of the Global Donation Network will require solving other challenges such as funding medical care in different countries and resolving differences in how kidney donations are legally and medically treated in each country. Ethical issues arise related to the perception that less developed countries could become sources of organs for well-developed countries and the concern that altruistic donation could give way to organ trafficking if not properly regulated.

3.1 WHAT IS TRUST AND WHY IT IS IMPORTANT

While AI increasingly supports analytics, humans are still very necessary for analytics implementation, and trust plays an important role. Trust is the ability to believe that another individual or group will follow through with a commitment [19]. In analytics, this commitment is that the math and the data are used in a way that produces an appropriate result. Whenever the concept or technique is beyond the understanding of another individual, then that individual must trust the AP.

In this chapter, we introduce two types of trust, swift, and conventional, that APs can use in building more meaningful relationships with stakeholders and customers. We look at trust in the mathematical model through the lens of interpretability and explainability and present some ways to help non-technical audiences understand the math behind a solution. Finally, we discuss ethical issues, both nefarious and unintentional, and share an example of an analytics team that navigated trust and ethical issues from many angles.

3.1.1 Types of Trust

The two types of trust are swift trust and conventional trust. Swift trust is formed over a period of minutes to hours. An oversimplification of this concept is the idea that first impressions matter. Do you look and act trustworthy during the first encounter with someone? Conventional trust, on the other hand, refers to the long-term and is built over a period of days, weeks, and months. It is reinforced by repeatedly fulfilling commitments on time. Conventional trust is very strong and is the key for the success of relationships that may span multiple years. How do we increase the likelihood of building these types of trust?

3.2 HOW TO BUILD SWIFT TRUST

Swift trust can be established quickly. The two important factors that influence the time it takes to build swift trust are defining your role and communicating clearly the promise of future collaboration [22].

3.2.1 Your Role Matters

It is important to define your role and make sure that everyone understands the scope of your involvement. From project to project, your role may change so it is helpful to communicate the change to make sure that everyone understands. For example, if your role in the project is either as an analyst or data engineer then that would be important to announce during any team meetings or discussions with decision-makers. In this case, being an AP would imply that your role would be developing and implementing a mathematical model and the role of the data engineer would be to connect the model to relevant data sources and perform extraction, transformation, and loading (ETL) of the data.

Do organizational titles help? To some extent. However, organizational titles may also lead to confusion. Perhaps your organizational title and role on the project are different. An extreme example of this is if an AP were the technical lead on a project and also the Director of Analytics at the organization. It is imperative that this AP establish that they are the project technical lead and reinforce that fact throughout the project. Due to the growth in the use of matrixed project teams with different roles for each project, there is an increased importance in using your project title as opposed to your organizational title to reduce confusion. In general, to decrease the time to build swift trust, only use your project role title during presentations or meetings that are directly related to the project.

3.2.2 Promise of Future Collaboration

After defining your role in the project, the next important factor in forming swift trust is creating the vision of future collaboration. The promise of future collaboration is simplified if there is a follow-on project or if this analysis is one of multiple phases in a larger project. Even if the project is a one-off effort, perhaps there is the opportunity to present the work at a conference, internal organizational showcase, podcast, or publish the results of the work online. Communicate this vision as early as possible in the project. Build it into the plan and reinforce this vision by repeating it as many times as possible during the project.

3.3   CONVENTIONAL TRUST

Building long-term trust takes time. Typically this is done by a series of small commitments that are achieved. In analytics, these small commitments are usually small scale analytic projects.

3.3.1   Defining Small Commitments

The defining features of small analytics projects are relative to the organization and the AP. It could be identified by a small dataset or the application of a straightforward analytics methodology. What is "small" at one company may be very large at another. The first step is to understand what a reasonable scope of effort and resources entails at your organization. Only after that determination can APs define what is small in terms of either the requisite data or the complexity of the methodology. APs build conventional trust by chaining together the successful implementation of many of these small scale analytics projects in a short period of time.

An important aspect of small projects is the ability for scaling to larger efforts. Meaning that a series of small projects will build upon one another to achieve a broader goal. For example, if the overall goal of the analytics department at a large retail company is to predict sales of a particular item using a machine learning model, then perhaps starting with just one store site is appropriate. If the organization operates worldwide then perhaps the scope of a small project for that organization would not cover just one site, but one city or region. Any unique features or parameters in the data for the small project sites, cities, or regions must be factored into scaling up this project to either a state, country, or even global level. Mapping out how the smaller projects will lead to the larger project goal before even starting takes both analytical experience and subject matter expertise. If the AP does not have either one of these then these characteristics may have to be found in other teams and/or team members.

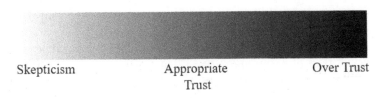

Skepticism   Appropriate   Over Trust
                             Trust

Figure 3.1 Spectrum of trust in the model.

3.3.2 Never Forget Larger Organizational Analytics Goals

Another consideration that impacts long-term trust is how smaller projects fall into the timeline of the larger organizational analytics picture. A scenario might exist in which many perceived wins with smaller projects result in an analytics team being overloaded with improving those analytical models and not having enough bandwidth to focus on the integrated analytics project. The pitfall of having inadequate resources increases the importance of not only mapping the smaller projects into the framework of the larger objective, but also the need to communicate the resource constraints of the smaller efforts. In this scenario, development of smaller models may need to be halted at a certain version if they fail to meet objectives or simply do not fit into the larger goals. The completion of a smaller model would allow for reallocation of resources to work on the larger project and manage expectations.

3.4 TRUST IN THE MATH MODEL

Things either work as intended or they don't. The emotional aspect of trust simply doesn't transfer to things, analytical models included. In Figure 3.1, we see the spectrum of perceived trust in the math model.

Achieving an appropriate level of trust relies on understanding both the benefits and weaknesses of a solution. The risk of entering the states of either skepticism or over-trust increases when one does not have enough knowledge about a solution.

The difference in the perceived level of trust of the solution between the AP and the decision maker is what you must focus on closing. Most people will assume that the difference is always that the AP holds a higher level of trust than the decision maker and that the AP needs to bring the decision maker up to their level. However, the opposite scenario can also occur, in which the decision maker has too much trust in the analytical solution or views an approach as a fix-all. In this situation, it is the responsibility of the AP to align their trust and clarify potential shortcomings or risks of a solution. The AP must make it a high priority to maintain the appropriate trust level for both themselves through thorough understanding of the model and with the audience by using effective communications.

3.4.1 Interpretability and Explainability

One of the greatest challenges for APs can be concisely answering the question "so, how does it work?". The answer requires translating a complex, multi-faceted mathematical model into terms that non-technical audiences can grasp, trust, and disseminate further. The requirement for a model to be human interpretable varies widely by industry, organizational culture, and circumstance. For instance, a medical model predicting patient outcomes may have much higher need for full human comprehension than a model recommending a song to add to a playlist.

Interpretability is an assessment of how well the model is able to be understood by humans. The decision tree is a clear-cut example of an easily interpretable model. Data is split based on various feature values, showing the decisions and cut-off values that lead to a classification. Each leaf node is a subset of the starting data at the root, which can be clearly defined by the decisions that generated it.

Many different ways exist in which you can assess the level of interpretability of a model. While it may be difficult to measure your model's interpretability from a full-scale perspective, you can assess the interpretability of a specific component individually. Breaking down high-impact model components into their simplest form allows someone to see it operate in real time.

How people experience and perceive advanced analytics models is crucial for their overall trust and adoption of the tools. In 2017 Google announced a new initiative called People + AI Research (PAIR) to bridge the gap between people and technology by focusing on user experience. One of the early products from the initiative called Facets allows users to explore either a high level view of their data or to do a deep dive into specific features. Often machine learning developers are not able to view their entire training data set because it is simply too large.

Consider for instance CIFAR-10, one of the most widely used collections of images for training machine learning and computer vision algorithms. Using the visual inspection features to review misclassifications, the PAIR team identified a mislabeled image [1]. The confusion matrix, shown in Figure 3.2, displays the actual image labels vs those that the model predicted. The diagonal shows accurate predictions and each pixel in the figure is actually an image, which can be visually inspected upon zooming in to better understand model behavior. Reviewing the images, they realized that one of the cat images their model had supposedly misclassified as a frog was, in fact, a frog. Their model had not made an error! Instead, the image had been mislabeled by a human labeler long ago. Engineers and researchers have used this dataset countless times without uncovering the error. Tools such as these show the importance of improving interactions between humans and machine intelligence.

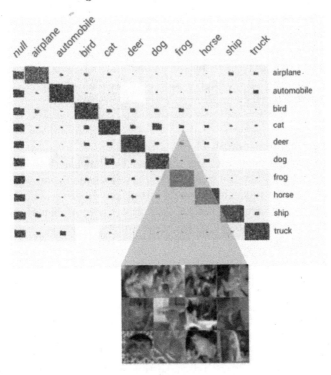

Figure 3.2 Google's PAIR facets.

Can cause and effect replace explainability?

The interpretability and explainability of models are important distinctions. At its core, explainability refers to a tangible relationship between inputs and model outputs. For a technical audience, explainability may include sharing commented code, concrete steps to analysis, and any other requirements for repeatability. For instance, if you are creating a stochastic model, this could involve setting a seed to ensure the same random number would be produced. For an audience of non-technical colleagues, explainability boils down to trust in the modeling solution. You can create this trust through a simplified, layman's explanation of the modeling approach combined with preliminary results that put the model into a real-world context.

While ensemble models and deep learning promise increases in model accuracy, these approaches can come at a cost to the explainability of the model. Approximations of black box models provide simplifications of a complex model, by using an explainable model that mimics the behavior of the black box model. While approximations give a better sense of cause and effect and allow us to assign functions to better understand behavior, they cannot truly explain what is going on under the hood of the original model.

A well-known example of an attempt to capture the behavior of neural networks is the use of saliency maps in image classification. Saliency maps show what part of an image a neural net is using to make a prediction, by highlighting the most impactful pixels. These often provide an "aha moment" by showing the AP that a neural net is in fact focusing on the part of the image on which the human mind would also place its attention. However, these maps have been criticized for their risk of introducing confirmation bias. Humans are more likely to believe a model is working if it aligns with their view of the problem.

3.4.2   Explainability with A/B Testing

Is there a way to mathematically explain why an analytics model produces a certain result? Perhaps Bayesian statistics can be a valuable tool to answer this question. Bayesian statistics have been gaining popularity in recent years. The popularity is largely due to advances in computing that make this kind of inference possible, but is also in part due to the explainability advantages of this method.

In frequentist statistics, a conclusion is reached using only data from the current experiment. Anyone who has attempted to explain the meaning of a p-value to a non-technical audience is familiar with the challenges of this method. Stakeholders want to know the answer to the question "is my hypothesis correct?", but the p-value answers the question "is my observed data unusual if the null hypothesis is correct?". In Bayesian inference, past knowledge, in the form of a prior, is combined with information from the current experiment to reach a conclusion. The result of Bayesian inference is a probability distribution rather than a point estimate. Organizations often have a wealth of valuable past data that can be utilized to help make decisions.

The century-long debate of frequentist vs Bayesian inference has made its way into one of the technology industry's most common tasks: A/B testing. In A/B testing, a randomized experiment compares two variants of some element of user experience. What do the two methods look like in practice? Imagine you are tasked with determining whether a company should change the color of their email sign-up button from blue to green. You have years of data on the click rate of the button in its current color, blue, and want to set up an experiment to determine whether the color change would make a significant difference in the number of sign-ups.

Under frequentist statistics, you would assume that the two colors for the button would have the same click rate and would study the difference with a unique experiment. While computationally fast, the frequentist test can take a long time in practice to collect enough data for a significant result. As a consequence, a common error in this approach is "peeking" at the test, which can create misleading results. When explaining the results to stakeholders, you are equipped with a p-value which does not give an indication of how much better one button color is than the other.

Next, let's look at this problem through a Bayesian lens. You begin with prior knowledge about the effectiveness of the blue button and update your belief based on new evidence as you experiment with the green button. By design, the analyst can peek at the test as evidence becomes part of the prior, which is quite valuable when explaining the results of the test to stakeholders. The AP can use this technique feature to more effectively convey the results of the analysis.

Journey to the bottom of the ocean and pain relievers: Does explainability matter?

We propose a thought experiment. You have been invited to take a journey to the Marianas Trench at its deepest known point of 10,984 meters (36,037 feet). At that depth, the pressure is 1,071 times the standard atmospheric pressure at sea level. Any small defect in your vehicle guarantees a certain instantaneous death. Two submersibles are available for your trip. Submersible 1 has made 100 journeys to the location experiencing no issues. How this vehicle works is not well understood. Submersible 2 has never been to that depth. However, there is extensive documentation and mathematical models confirming that it should make the trip without any issues. Which one do you choose?

If analytical explainability is the decision metric then Submersible 2 is the dominant choice. However, humans tend to choose Submersible 1 over Submersible 2. Why is this? Of the many possible explanations the most likely one is that Submersible 1 has more of what humans want to see in explanations. That factor is observable real-life success. Humans weigh evidence seen with their own eyes higher than mathematical equations and outputs produced by a computer. Relying on physical observations have ensured the evolutionary success of humans for tens of thousands of years.

One real-life example of humans relying on products they don't fully understand is Tylenol (acetaminophen). It has been around since the 1940s because of its utility in treating pain and reducing fever, among other things. Tylenol relieves pain by blocking the body's production of a substance called prostaglandin, which is associated with inflammation. How it does this is not understood: however, that has not reduced our collective trust in the drug. Tylenol is the most commonly taken analgesic worldwide and is recommended as first-line therapy in pain conditions by the World Health Organization (WHO) [13].

The impact of this glitch in the human evolutionary processing system means that the mathematical explanations of analytics solutions have an uphill battle when competing with solutions that have a track record supported by historical performance even if the cause of that historical performance cannot be explained. It is important that APs understand this concept and consider its effects when developing explanations of their model results.

3.5   ETHICAL ISSUES

The two major ways that APs can use their skills unethically are with or without intention. The first way, nefarious use of analytics, refers to when APs use their skills to deceive or mislead. The desire to intentionally mislead can come from the AP themselves or from a supervisor or stakeholder who asks the AP to ensure a certain result. The second way is much more common and occurs when the person uses analytics inappropriately but does not realize it.

Nefarious use of analytics is the deliberate manipulation of data, methods, and processes to produce results that are deliberately misleading or false. Nefarious use of analytics includes intentionally entering incorrect numbers into a spreadsheet or report, withholding important data from the wrong people, purposely not using reliable models when they are available, or abusing data mining to find evidence that supports a desired conclusion. The goal of nefariously using analytics is usually to get an advantage over others. The harm occurs when people make decisions based on the misleading or manipulated information instead of basing their decision on complete and accurate information. There is also additional damage if people begin to distrust the analytics team, which could make it more difficult for fellow APs to use their skills ethically in the future.

As the scope and ramifications of nefarious analytics are well-known, we'll focus on the risks of unintentional ethical issues for the remainder of this section.

3.5.1   Detection of Unintentional Ethical Issues

Unintentional ethical issues arise when APs lack understanding of the full implications of their analysis. In order to better understand how APs can use data incorrectly, we'll look at the example of implicit bias.

One of the pillars on which many large technology companies operate is that they will not discriminate against certain groups or minorities. However, their products might introduce discriminatory practices without knowing it. One way this may occur is by using AI to predict the future decisions of a person based on past actions. For example, data may show that individuals who have defaulted on their loans in the past are more likely to do so again. Someone with no credit history might then be seen by an algorithm as less likely to default because they lack this characteristic. Under these assumptions, people who were previously excluded from loans might now be getting accepted due to having no credit history. While the mortgage industry has claimed that their algorithms control for factors that may introduce bias, a 2019 study showed that nationally American lenders were 40 percent more likely to reject Latino applicants and 80 percent more likely to turn down Black applicants for loans than similar white applicants [28].

Another example of well-intended AI gone wrong is its use to eliminate unconscious gender and racial bias in the hiring process. The idea is that an employer could use machine learning software to filter out resumes based on objective criteria such as education levels and job experience. However, the software would also include implicit bias in the way it judges resumes as it will select those that match the "good" resumes on which it was trained, reinforcing existing biases.

3.5.2   How to Mitigate Unintentional Errors

Since the trust in the AP may act as a proxy for the trust of the supporting math, the AP has the responsibility to at a minimum effectively communicate the model's strengths and weaknesses through the lens of the business problem. By documenting any assumptions, issues or questions about the data, possible blindspots with the model, and potential areas for bias in either the data or the algorithm, the AP can use this transparency to communicate any weakness. The documents should provide another AP or organization the opportunity to check the work and access if there are any unintentional errors. The reviewing organization could be either an internal or external group. Periodic review for these errors is also recommended due to the ever-changing standards in this area.

3.6   REAL-WORLD IMPACT - PREDICTIVE POLICING

Predictive policing has brought conversations about the role of trust in analytics into the public sphere. The practice of predictive policing aims to make law enforcement more effective and fair through the use of big data and predictive algorithms. One of the most successful such systems is the New York City Police Department Domain Awareness System (NYPD DAS), a digital surveillance network that evolved from counterterrorism to general policing applications. In 2017, four years after its department-wide implementation, the overall crime index in New York City fell by six percent, with an estimated annual savings of $50 million resulting from the tool.

The NYPD DAS is supported by data from license plate readers, cameras, sensors, and 911-calls. Pattern recognition methods using license plate location data allow officers to forecast the location of vehicles on their watch lists. Algorithms also help determine resource allocation by automating the creation of hot-spot maps, which place officers in optimal locations to prevent future crimes. Another popular policing platform in the US is Pred-Pol, developed by the Los Angeles Police Department (LAPD) and used in precincts across the country. Pred-Pol uses the date, type, and location of individual crimes to predict the occurrence of future ones. The LAPD announced the end of the use of the platform in 2020 [29].

Critics of these methods have called the DAS platform "Orwellian" and compared the pre-crime methods of Pred-Pol to the novel Minority Report by Philip K. Dick [26]. Considering the high impact of these tools, their developers needed to gain buy-in from not only decision-makers in their departments, but also the end-users and the community in which they operate. Each of these stakeholders would have different concerns about the usefulness and nature of these tools. For instance, the DAS was met with reluctance from officers on the effectiveness of its supporting algorithms, as they had seen unsuccessful tools in the past and doubted its ability to replicate human processes:

"NYPD officers also tend to be skeptical of analytics solutions that run counter to their instincts. That is, if we deployed a black-box solution, our officers would likely not trust or use the outputs. For this reason, we make sure to employ techniques for which the underlying records influencing the outputs can be made visible, thus enabling officers to see the algorithm's drivers." [24]

Bringing police officers into the development cycle helped the NYPD to create a transparent solution that officers would trust. In both New York and Los Angeles, community stakeholders and civil liberties groups presented concerns on privacy and model bias. Predictive policing tools rely on historical crime data to determine where officers should be placed at a given time. If crime rates in communities of color are historically high and these areas are historically extensively policed, they will continue to be so under the basis of predictive policing. Despite anonymization, data driving these algorithms has the potential to mirror existing inequalities and reinforce systemic bias.

In a sensitive case such as this, the two best tools for building trust are transparency and stakeholder participation. By representing the diversity of the department and community in the model building process, modelers can begin to identify and mitigate bias inherent in the data.

EXERCISES

3.1 How do you establish trust upon first contact with a new stakeholder?

3.2 Provide an example of a recent analytics project you participated in and consider how you could have defined your project role and promise of future collaboration to the customerclient?

3.3 What special considerations for establishing trust apply to predictive policing?

3.4 What is the relationship between trust and transparency when it comes to analytics? How can you maintain trust when you cannot be 100 percent transparent about an analytics solution due to project limitations?

3.5 Identify the stakeholders that the analytics team developing the kidney donation model would have to convince of the model's effectiveness?

FURTHER READING

Hawley, K. (2012). *Trust: A Very Short Introduction (1st edition)*. Oxford University Press.

Sigma Xi (2000). The Scientific Research Society. *Honor in Science*. Sigma Xi, the Scientific Research Society.

CHAPTER 4

Communication

> "If you can't explain it simply, you don't understand it well enough."
>
> Albert Einstein

T HE UNITED NATIONS World Food Programme (WFP) is the largest humanitarian organization supporting food security worldwide. Responding to ongoing food security issues as well as sudden challenges posed by climate emergencies and conflict, the WFP operates in a variable environment. In addition, they grapple with a low budget and the need to meet expectations from involved stakeholders, including individual donors, governments, and taxpayers.

Food delivered to sensitive areas must meet time and quantity requirements, as well as providing a nutritional balance to sustain a healthy population. For operations research analysts at WFP, this means combining supply chain decisions with decisions surrounding the nutritional breakdown of food delivered. In collaboration with researchers from Georgia Tech and Tilburg University, the WFP developed a tool called *Optimus* that achieved this by integrating multiple decisions into a single model [30]. The solution improved operations by changing the siloed nature of supply chain decision-making into a centralized model. However, after several years of use, the tool was reevaluated for its time requirements and the need for an analyst-in-the loop. The model could not be run without a

DOI: 10.1201/9781003204190-4

Figure 4.1 Food for distribution. v2osk, 2016.

user with a background in operations research. WFP employees could see the final solution, but questioned how the model came to that decision, and why it selected certain items to include in the food basket.

In the following months, the team focused on automating the data pipeline and optimizing the user design so that anyone in the organization could use the tool. The redesigned model allowed users to interact with the decisions the model made, exploring what items in the food basket, as shown in Figure 4.1, had been swapped out in different scenarios. This involvement in the modeling process built trust in the final solution. For beneficiaries, the team also used their results to clearly link funding levels to nutritional targets, empowering stakeholders to understand the trade-offs in the business problem.

APs working on this project needed not only to solve a complex mathematical problem, but also to share actionable results with two very distinct audiences: employees working in the field and donors. Experienced WFP employees working in the field have rich subject matter knowledge and likely their own mental models or heuristics regarding supply chain requirements. However, they likely do not hold advanced math-based degrees. For WFP, donors could be government representatives, corporations, or even just individuals who are in support of the mission. Analysts needed to speak two different languages to reach these audiences, and present model results in terms of the metrics that each audience deems important.

In this chapter, we explore facets of communication that determine how analytical information is perceived. We cover how to build rapport and make the audience care about your message. In contrast to the traditional focus on communicating with non-technical audiences, we also share some key points for effective communication within technical teams. Finally, we look at some examples of communication in practice and how to use data visualization to tell a story.

4.1   THE GOAL OF COMMUNICATION

The goal of technical communication is to truthfully represent the data and equip the end user with the information they need to make a decision. Communication is a critical skill to analytics success: from requirements definition with stakeholders, to explaining a method to technical colleagues, to presenting a final solution to clients. Each of these scenarios requires a slightly different communication style, generally refined through experience. Communication is an opportunity to translate data into stories and make valuable analytics accessible to a broader audience. How analytics is communicated affects how the work of the AP is interpreted and ultimately, whether or not resulting recommendations are put into action. For the WFP, this level of communication was achieved through direct interaction with the model, representing results in the language of the user and stakeholder. This led to not only implementation of the model, but also a cultural shift toward acceptance of data and analytics.

Before engaging with a stakeholder, consider the following questions:

Why am I here? Consider the background story of what brought you to this location, date, and time. The context of the reasons and the events leading to this moment are important.

Who am I speaking with? Do the research and figure out what the audience's background is to conceptualize in what context they may be framing both the business problem and your analytical solution. It is very important to know what level of explanation to provide.

What do I want to accomplish today? Identify the objective of the conversation or presentation. You can't always provide a college course in every presentation.

What is my overall mission? The final goal of the engagement may fit into a larger project or business context, which is important to understand.

Perhaps this session is a project kick-off, or perhaps it is part of a recurring series. The context impacts the amount of information provided and what the audience should understand in the solution space after speaking with you. Communication goes both ways. The AP must observe the audience and perceive whether they are understanding the message. Your message or explanations may need to be dynamic, responding to the areas in which your audience needs more clarification or holds more interest.

4.2   LAYING THE GROUNDWORK FOR SUCCESSFUL COMMUNICATION

Rapport enables us to put others at ease and have more meaningful interactions with them. It also increases the possibility of hearing our message. The ability to establish rapport with other people is an essential skill in many situations. The following are some guidelines on establishing positive rapport, as well as how to maintain it afterward.

4.2.1   What Is Rapport

Rapport is communication built on trust. It requires that we see things from the other person's point of view. When they feel understood and respected by us, it makes them more likely to listen to what we have to say.

> **Establishing rapport.** In order to establish rapport with someone, you must first establish trust. You can do this by using nonverbal signals such as open body language and maintaining eye contact. The nonverbal signals

will show the other person that you are receptive to them and do not have any hidden motives. You can also use verbal cues from active listening, such as paraphrasing what they say in your own words, to demonstrate that you are listening without interrupting them or imposing your own views. Once you have built trust, it will be far easier to establish rapport with the other person at a deeper level.

Maintaining rapport. After having established rapport with someone, you should continue to show that you are open and trustworthy in your interactions. You should not try to change anything about yourself; instead, act the same way you would normally. If your rapport is working well, then they should feel comfortable around you and continue to be open with you.

Starting a conversation. A simple way to build rapport is by starting a conversation. This doesn't have to be anything formal; in fact, it works best if it feels natural and unforced. If you know the person well, then ask them about themselves and their interests. If not, try to find something that you can relate to in order to get things moving; for example, remark on something that is happening in the environment around you or mention a recent event in another part of your life where there was mutual involvement.

4.2.2 Factors to Success Discovered through Research

A 2008 study on rapport [17] in a retail environment used surveys of both employees and customers to analyze the most effective behaviors to create rapport. The following categories were used to classify the employee behavior with a customer.

Uncommonly attentive behavior, atypical actions
These employee actions were perceived by the customer as the employee going out of their way or "above and beyond the call of duty" to satisfy the customer's request. In analytics, an example of this is an AP providing additional analysis or data focused on a closely related business question to the original problem.

Connecting behavior, pleasant conversation The behavior in this subcategory covered initiating a conversation to engage in enjoyable interaction with the customer. The subject of the conversation did not focus on the business transaction and is better known as "small talk." The use of this behavior in an analytical situation might take the form of discussing the weather or perhaps the latest big game for an audience member's favorite sports team before a presentation even begins.

Information sharing behavior, imparting knowledge An employee shares their expertise in certain areas where the person has insight or knowledge that the customer does not have. In the analytical context, an example of this would be if the AP had performed prior research into a subject closely related to business questions the decision-maker may ask during the brief. If the AP provides insights to those questions during the brief when asked, this would build rapport at a higher rate.

4.3 SETTING THE CONTEXT

In the Indian parable of the Blind Men and the Elephant, a group of six blind men approach an elephant and try to determine what the animal is by feel. One man touches the elephant's leg and proclaims it is sturdy and large like a tree trunk. Another grabs its tail and notes its similarity to a rope. Yet another feels the elephant's tusk, sharp, and smooth, and believes it is like a spear. As each man touches a different part of the animal, they make different assumptions about it. Independently, their conclusions are all incorrect. However, cumulatively, they can create an accurate description of the animal. An AP independently approaching a business problem with only raw data is not too different from a blind man approaching only the trunk of an elephant: they will likely characterize a part of the problem correctly but will miss out on the broader picture. This pitfall is why communication is essential in the practice of analytics. Active communication before, during, and following the analytics project grounds the results in the context of the business problem.

The listening side of communication comes into play in establishing the project context. Understanding the setting of an analytics project will help you to determine how the effort fits into the larger business picture. The setting includes how it will be used and by who, what support is available, and why the outcomes of the project are important. Some project environments will be subject to a high degree of variability or incur significant effects from outside influences. Understanding this environment: how it impacts the data quality as well as the team dynamic is essential. Observing, listening, and asking targeted questions are the best tools to gain insights into how the analytics problem at hand fits into the larger business environment.

4.4 MANAGING EXPECTATIONS

With the recent burst in popularity of advanced analytical methods, more and more people are becoming analytically literate. Machine learning and optimization have become part of standard business jargon. However, this does not mean that everyone understands how these methods work or in which scenarios they are applicable.

As an AP, how do you deal with a boss or stakeholder armed with just enough knowledge about these methods to be dangerous? What if a client approaches you and shares a successful use case of deep learning to fight fraud by a big tech company? The client suggests that you also apply this method to their transactions, but you are wary that there simply isn't enough data as their operations are considerably smaller. Appraise the recommendation thoughtfully and honestly share the strengths and limitations of applying a certain method to your problem. Often nuances exist that impact application to specific scenarios. If another approach is a better fit for the problem at hand, share this approach as well, in combination with why it could be a better fit.

Remember, it is your job to communicate what approach will solve the problem best within the bounds of proper statistics. A scenario like this is a perfect opportunity to build trust which we covered in chapter 3. Individuals new to advanced analytics may view these approaches as a fix-all or place too much faith into their solutions. When a stakeholder is knowledgeable about an approach or very invested in the technical details of a solution, create an opportunity for them to actively engage with the analysis. By delivering results in a transparent way, stakeholders can feel involved behind the scenes and in turn, have a better understanding of the analysis. In implementation, a transparent solution could incorporate levers or buttons that allow a viewer to test scenarios and understand cause and effect within an analytics model.

You've probably heard the phrase "underpromise, overdeliver" as a method to manage expectations. For APs this can be particularly relevant, as often you can calculate a range of impact or improvement you expect from a project. If you share the absolute best case scenario, stakeholders will likely remember and judge the success of the project against that metric. If you share a lower- to mid-range of improvement, they will be happy when you meet it and ecstatic if you go above and beyond.

With experience and relationship building also comes the ability to anticipate client needs. As you get to know a stakeholder, the problems they are most interested in solving will become clear. Additionally, once you have completed a project with a stakeholder, you can gather valuable feedback about what went well or could have been improved. Some individuals will place more weight on numeric outcomes, others on the look and feel of a project. Anticipating needs is an aspect of communication that must be learned by practice and reinforced by observation.

4.5 MAKE THE AUDIENCE CARE

Math Anxiety (MA) is a documented psychological condition that is defined as the feeling of tension and fear that accompanies math-related activities [31]. MA creates a barrier between the AP and the audience that begins with the nervousness that the audience is likely to feel even before the meeting. Audience members may even experience the same anxiety that they felt before a test in a high school or college math exam before an analytics presentation. Considering this condition, reference the business question to be answered in the title of the presentation, instead of leading solely with analytics, model, data, or anything related to math. Overcoming this anxiety is a key objective for any explanation in analytics as it reduces the barrier to understanding and the potential for meaningful conversations.

The business question provides the anchor for the entire explanation. For example, if the business question is how to schedule city buses more efficiently to save money, transport more people, and reduce the variation in schedule variance, then that should be the focus of the brief. Any data or methodology material presented should support answering this question

After getting buy-in on the analytics addresses the business question, the next step is creating an environment in which people care. Emotions are individual and everyone reacts differently. Passing roadside litter, one person sees a piece of trash by the side of the road and cares enough to pick up the trash and throw it away. Another person may care so much that they are motivated to organize a road-side cleanup effort or propose installing a sign that discourages littering. Still another person may see the same piece of trash, not care, and do nothing. The difference in these three scenarios is how the person perceives the piece of trash.

The trash could be perceived as just trash or it could be perceived as a sign of greater harm to the environment. Although no one can control this perception for each individual, the presenter can lead the audience on a story that provides context. In the case of our example, a presenter could show a picture of one piece of trash on the side of the road and then walk the audience through the lifecycle of this rubbish. Perhaps it makes its way into local waterways and decomposes into microplastics. Microplastics disrupt the ecosystems and even make their way back into the food system. The story increases the probability that more people would care about this issue.

4.5.1   Explaining Complex Topics at the Right Level

We have all sat through a presentation we wished we could leave. These challenging talks share a few things in common: a muddied message lost in a complex display, information presented at too high of a technical level, or simply information that bears no relevance to the audience. Great presentations, on the other hand, hold the audience's attention by clearly conveying their value to the audience through a story that engages the audience, shared knowledge that can inform decision-making, or motivation for future action. While recognizing a great technical presentation is easy, crafting one, especially under time limitations and pressure, can be a challenging task.

All data tells a story, which APs uncover as they work through an analysis. For example, this underlying story could show a change over time, trends by region, or show familiar patterns that allow us to detect anomalies. Whatever the story, share this with the audience to help them understand the results from the same perspective that you do. The information shared could be assuring, in line with their picture of reality or enlightening, in contrast with their expectations. Speakers can ensure there are no comprehension barriers by explaining potentially unfamiliar topics and terms being used in the presentation upfront. Some topics, such as machine learning techniques are impossible to explain in depth to a non-technical audience in a short period of time.

New concepts can be made clear with the use of analogies, which help us to understand something new in the framework of something familiar. Empirical studies in cognitive science have shown how analogs can effectively help learners jump into a subject area in which they have limited to no prior knowledge [33]. For example, a common analogy used in introductory statistics classes is the comparison between hypothesis testing and a criminal trial. Students can easily associate the idea of "innocent until proven guilty" with a null hypothesis. When explaining a concept with analogy, it is important to clarify which elements of the example are analogous, and which fundamentally differ [27]. For instance, over-comparison of deep learning to the human brain has led to some misconceptions from the general public about the field.

From genetic optimization to swarm heuristics, several advanced analytics methods were inspired by nature. Beyond their effectiveness, many of these methods are also simple to explain to a non-technical audience due to familiarity with the natural phenomena that inspired them. For instance, ant colony optimization, a method of finding paths through graphs, brings to mind familiar, memorable imagery. Most people can recall observing the behavior of an ant hill in their childhood. Learners can then use this imagery to grasp how the method is inspired by the ants' communication through pheromones that inform other ants of a favorable path. For APs, the ability to bring a complex analytics topic into the imagery and understanding of something familiar is a valuable asset.

Communication is a two way street. It is important for both parties to listen and talk so they can really understand one another. APs need awareness of when the explanation of either the model or the data has become unclear to the audience. If this happens, time needs to be spent allowing the audience to ask questions and get clarification. It's also important to leave time during a presentation for discussion. Extra time provides the opportunity to get feedback from your audience that can help you present your point better or identify subjects that need more clarification.

4.6 ATOMIZING KNOWLEDGE

When a pastry chef creates a flakey dough, they don't simply throw a full stick of cold butter into the flour and expect it to integrate. Instead, they cut the butter into small pieces, gradually working it into the flour. If you regularly collaborate with a set of stakeholders, atomizing knowledge can be an effective way to build technical knowledge over time. Through atomized learning, small, memorable pieces of information about a topic are shared at a time. As the audience gains comfort with topics and sees them in practice, new topics can be added into the mix.

In Figure 4.2, the analytical method being atomized is Natural Language Processing (NLP). NLP can be atomized into components of sentiment analysis and topic modeling. These components can be further broken down into smaller and smaller topics (as shown in the figure) which then become individual teaching points. In your initial presentation using NLP, one slide could be dedicated to teaching the concepts of stop words. The next presentation could present the topic of stemming and so on.

Decision-makers need a basic understanding of the strengths and weaknesses of analytical approaches, but APs don't have the luxury of an entire college semester or two to teach advanced mathematical concepts. Playing the role of an educator is an excellent way to build trust. You can explain what you found out about the business problem, how you got there, and why it matters.

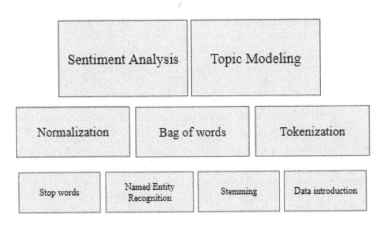

Figure 4.2 Atomizing NLP.

TABLE 4.1 Levels of Understanding

Scale	Level	Description
1	Ignorance	A person does not understand or know about a subject
2	Knowledge	A person isaware of a subject
3	Comprehension	A person understands and knows about something, and can also explain it to others
4	Application	A person understands and knows about something, and can use it in practical ways
5	Integration	The highest level of understanding, where a person not only understands and knows about something, but also sees the connections between it and everything else

4.7 LEVELS OF UNDERSTANDING

Table 4.1 introduces five levels of understanding, which the AP must assess for both themselves and their audience. The AP should be at a level four or five for any topics that directly impacted their work on the development of the model. They should have at least a level three, comprehension, of aspects of the model with which they were not directly involved. When the audience is at a level one, ignorance, then atomizing the knowledge could be a useful way to move them up the scale.

Complex concepts can be broken down into very small bite-sized chunks which can be delivered during periodic update meetings over time. Done correctly, the AP may start setting up a large concept required to understand the analytical approach weeks or months in advance.

For example, picture your team using the unsupervised ML k-means clustering methodology as the analytical approach. The team is briefing the decision-makers every two weeks over the course of a six month project. The team could break the k-means approach down into bite sized-chunks. One chunk might be identifying the data to cluster, scaling the data, a one slide overview of least mean squares distance determination, and determining the number of clusters. If all of these concepts were covered in one brief then the non-APs would more than likely feel overwhelmed and uncomfortable with the approach. Provided over a period of time, the non-APs should hopefully gain an incremental increase in the understanding of the concept and at the end of the project have the appropriate trust for the technique. The method also enables the non-APs to ask higher quality questions and actively participate in future solution development.

4.8 GATHERING INFORMATION ON PROCESSES AND DATA

Gathering data when carrying out design, development, and validation, is essential. You can use several different techniques for data collection, each with its own strengths and weaknesses. The most common methods are surveys, interviews, focus groups, and workshops. Let's take a closer look at each of these.

4.8.1 Surveys

Surveys are an important tool for collecting data. They allow you to collect information from a large number of people in a short amount of time. When designing a survey, it is important to first identify the research question or questions that you are trying to answer. The questionnaire should be designed to collect data that will help you answer the questions you have identified. It is also important to pretest the survey to make sure that it is effective and that the questions are clear and easy to understand.

When creating a survey, it is important to first identify the research question or questions that you are trying to answer. The questionnaire should be designed to collect data that will help you answer the questions you have identified. It is also important to pretest the survey to make sure that it is effective and that the questions are clear and easy to understand.

Once the survey is designed, the target population is determined. The target population is the group of people that you want to take the survey. It is important to make sure that the target population is defined in a way that allows you to collect data. For example, if using an online survey then make sure that the target population has access to a computer or mobile device. The next step is determining the sample size. The sample size is the number of people who will take the survey. It is important to make sure that the sample size is large enough to get a good representation of the target population. The sample size also depends on the type of data that you are collecting. For qualitative data, you may need a smaller sample size than for quantitative data.

The strengths of using a survey is that it can be used to collect large amounts of data in a small amount of time. The weaknesses include a low response rate and survey fatigue if the technique is used too frequently. Surveys have other constraints such as rules and policies on how they are administered, who can participate, and if the survey takers are kept anonymous that vary by organization.

4.8.2 Interviews

Interviews are a common method of data collection used in qualitative research. They involve interviewing people one-on-one to gather information on their thoughts and experiences. Interviews can be either structured or unstructured and can be audio or video recorded. An advantage of interviews is that they allow for in-depth exploration of a topic. However, they can be time-consuming and expensive to conduct.

When conducting interviews, it is important to ensure that the questions are relevant to the research question and that they are asked in a way that does not bias the responses. Interviews can be analyzed using qualitative methods such as content analysis or thematic analysis.

Some tips for conducting interviews:

- Plan out your questions in advance and ensure that they are relevant to the research question.

- Make sure you are familiar with the topic of the interview and can ask questions that probe further into the respondent's thoughts and experiences.

- Be respectful and attentive during the interview.

- Ensure that the interview is audio or video recorded. The recording will help to ensure accuracy and allow for later analysis.

- Make sure to thank the respondent for their time once the interview is completed.

Interviews can provide valuable insight into people's thoughts and experiences. When conducted correctly, they can provide rich data that can be used to answer research questions. By asking relevant questions and being respectful of the respondent, interviews can be a useful tool for qualitative research.

4.8.3 Focus Groups

Focus groups are a type of qualitative research methodology that involves gathering a group of people together to discuss a topic. Focus groups can be used to gather both qualitative and quantitative data. One advantage of focus groups is that they allow participants to interact with each other, yielding more insightful data.

Here are a few different types of focus groups:

- Topic focus groups: In these groups, participants are asked to discuss a specific topic or issue.

- Thematic focus groups: In these groups, participants are asked to discuss a theme or topic that is relevant to them.

- Open-ended focus groups: In these groups, participants are encouraged to respond freely and help to guide the conversation.

- Closed-ended focus groups: In these groups, participants are asked to answer specific questions about a topic or issue.

- Depth interviews: These are similar to focus groups, but involve one-on-one interviews with participants.

When selecting participants for a focus group, it is important to consider who will be most likely to contribute useful information. A few different ways to select participants are:

- Random sampling: In this method, everyone in the population has an equal chance of being selected.

- Stratified sampling: In this method, the population is divided into groups (or strata) and people are selected from each group at random.

- Cluster sampling: In this method, clusters of people are randomly selected, and then individual participants are chosen from within the clusters.

- Quota sampling: In this method, a certain number of participants from specific groups are chosen.

- Convenience sampling: In this method, participants are selected based on convenience (for example, people who are available at a specific time and place).

Once the focus group is assembled, the moderator must ensure that everyone has a chance to speak and that the discussion remains on topic. The moderator should also be aware of potential biases, such as group dynamics and the influence of dominant participants.

Focus group strengths include being able to receive feedback and input from multiple participants taking the same amount of time as just a few one-on-one interviews. The drawbacks to this method is scheduling a time that works for everyone selected to participate, group dynamics such as junior participants not wanting to share answers with senior participants present, and requiring an experienced moderate to get the most from the event.

4.8.4 Workshops

Workshops are typically one to multiple day events. Workshops provide an opportunity to bring a target group to one location and participate in multiple elicitation opportunities. An example of using this method is to bring the users of a deployed analytics tool together. Over a few days, updates on new versions of the analytics tools could be presented, training conducted, one-on-one interviews held with experienced users, and a focus group with new users. Attendees could take a survey after viewing the presentation on the new product features to determine interest levels.

The strengths of workshops are that attendees are focused on the event for the duration. The challenges with workshops are large effort in planning and the lead time before the event. Unless the event is online, there are the logistical issues of attendees having to plan and execute travel to the venue.

4.8.5 Combination of Techniques

As with the workshop, all of these techniques could be used in combination with each other. You could use a survey to determine the best candidates for one-on-one interviews or follow up a focus group a few months later with one-on-one interviews. Numerous ways exist to combine these techniques to optimize the information in both quality and quantity over a given period of time. Don't restrict yourself to just one if time permits.

4.9 COMMUNICATING WITH A TECHNICAL AUDIENCE

While most effort is understandably placed on communicating technical concepts to non-technical audiences, alignment, and collaboration within technical teams is equally important. Technical communication can take several forms: collaboration on a project, peer review, or simply sharing a relevant technical approach to an interested audience, as at a professional conference. Similar to non-technical communication, effective technical-to-technical communication requires knowing your audience and the appropriate level of detail to share. The mistake that APs commonly make is assuming a technical environment means it is appropriate to share every minute detail of their process without distilling the analysis.

4.9.1 Sharing Your Analytics

While at a team meeting the goal is to use feedback to improve your research, at a conference the audience generally has interest in how your project could inform their own work. A team meeting will be more informal and conversational with questions throughout and your presentation should be structured to accommodate and encourage this. Regardless of the avenue, technical communication allows you to showcase what elements of an analysis will be of most interest to the technical stakeholders, while soliciting feedback from those who have expertise in either the data or methods you are employing.

It can be exciting to share technical details without feeling the need to translate or embellish a presentation. However, not putting effort into the look and feel of a presentation for technical audiences can be detrimental to your message. The same rules of effective storytelling and information sharing apply in this case. The use of figures and refined graphics will have a major impact for visual learners in a technical audience.

4.9.2 How to Share the Right Level of Information

While it may not be feasible or appropriate to share step-by-step methodology in a technical-to-technical presentation, many technical audiences may want to review details of your analysis on their own time. A helpful asset for this situation is the use of documentation. Documentation, in the form of comments or user guides, is key to guiding technical stakeholders through your analysis. It provides a window into your approach to a problem and allows interested parties to interact with the analysis without reading every line of code. Similar to the process of writing, there is always a creation stage and an editing stage to coding. If you are coding by yourself, readability can feel less important than functionality. However, keep the audience in mind, whether collaborators or future users, as you refine code.

As with non-technical communication, structure your message around the audience goals. If you are at a conference, where attendees may have a different use case for your presented methodology, share some context-agnostic tips for implementation. For instance, share the benefits and drawbacks of the method you applied. If you are sharing an analysis with a coworker who specializes in data management, prepare specific information about how you manipulated the data. By stepping into the shoes of the audience, you can both provide a more interesting presentation and be more prepared for the types of questions that may come up.

4.10   GUIDING THE CONVERSATION

Unfortunately, what is most interesting to an AP is not always what will be most interesting to their audience. Balancing sharing the right amount of methodology explanation while also explaining how the results fit into the business is a delicate process. You could easily waste an entire meeting diving into technical details or getting sidetracked by specific questions related to the project context.

Concrete, achievable goals for a meeting not only allow an AP to focus on their own content to share, but also to ensure the conversation stays on track. For instance, instead of creating a presentation to state the results of a project, use the time to make a data-backed recommendation for action. In lieu of simply sharing an exciting new model, use a focused presentation to gain buy-in for a particular method, showing tangible examples of how it could impact current processes. Always plan time for discussion to get feedback on analytical questions during presentations and meetings.

4.11 USING DATA VISUALIZATION EFFECTIVELY

4.11.1 A Picture Is Worth a Thousand Words

Any discussion of communicating analytics would be incomplete without considering data visualization. Visualization allows us to understand complex relationships and patterns in data more quickly and effectively than by looking at raw data. We use data visualizations to make quick decisions daily. From viewing Doppler radar on the weather channel to measuring progress on a fitness tracker, data visualizations influence our day-to-day activities.

Data visualizations usually fall into one of two categories: exploratory or explanatory. Exploratory visuals are helpful as you familiarize yourself with the data and can be useful to share insights with other analysts. Scatter plots and distribution graphics are two of the most helpful exploratory visuals to tease out trends and relationships in data. Explanatory visuals, on the other hand, are those that you would share with a stakeholder as they display something you directly want to communicate.

A strong visual is often more memorable to an audience than a stand alone statistic or numeric outcome from an analysis. In many ways data visualization is a skill quite distant from the AP's usual toolkit: beyond statistically sound inputs, it requires command of aesthetic elements and an understanding of what will be most impactful to the viewer.

In user experience design, the phrase "don't make me think" refers to making products readily accessible to the user [23]. The user shouldn't spend ages contemplating how to use a webpage as it will waste their time and make them less likely to return to the product. What does intuitive design look like for analytics? It can apply to how applications are developed, how presentations are structured, and how data is visualized. For instance, consider you are making a dashboard and you are including a plot of how a certain sustainability rating changes over time. Users do not work with this rating often and may forget whether lower or higher values are "good." Instead of labeling the axis "Sustainability Rating," you could instead label it directionally, indicating "Improving Sustainability Rating." The simple change makes the visualization more accessible and removes the extra step of the user researching the rating to understand your graphic.

4.11.2 What Is Effective Visualization

Effective visualizations provide the viewer with a complete, concise picture of the analysis. As you select a visualization, consider the relationship you would like to showcase: for instance, distribution, correlation, ranking, or parts of a whole. Each element should be carefully considered for what information it represents and the value it brings to the visualization. Furthermore, the way data is displayed through visual encodings, can greatly impact how a number or relationship is perceived. For instance, the visual encodings of position and length have been shown to be more precisely interpreted than angle and the intensity of color [15]. The visual encoding precision is why a bar chart is often easier to read than a pie chart. The balance between streamlined graphics and chart clutter can be difficult to navigate. While as a general rule, simple is usually better, simplification can also be taken too far at the expense of valuable information.

4.11.3 Employing Data Visualization for Change

What data shows is almost always just one piece of a broader picture. Placing analytical results in the context of the business question, project, or broader trends helps to show its importance. In this manner, data visualization can be employed as a storytelling device. Visual elements can display space, movement, and change over time. An AP can guide the audience's attention to particular parts of the story with the use of graphical elements. For example, color is a useful tool to emphasize or deemphasize parts of a graphic.

Text used as a label is an often overlooked way to inject graphics with context. Supporting information can be directly included in a graphic with the use of subtitles, captions, and annotations. Including a direct call to action in the title or subtitle ensures the viewer takes away the intended message from the graphic and knows how to put any recommendations into practice. Lead the audience to the conclusion you want them to draw from the visualization, and when appropriate, what action they should take based on this conclusion. When these elements are included directly on the graphic, you reduce the risk of the graphic being lifted from its presentation and shared without proper explanation.

Figure 4.3 provides an example of an annotated data visualization that tells a story. Without text, the graphic would simply

Kidney transplants on the rise

While kidney paired donation became legally possible in the US in 2007, it took several years to reach adoption. It has since increased the number of possible living donor transplants.

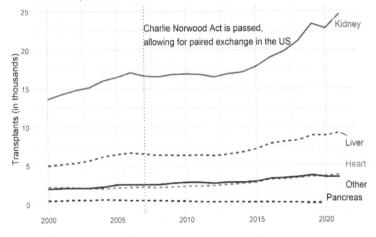

Source: Organ Procurement and Transplantation Network (OPTN)

Figure 4.3 Example of text annotation on a chart.

display the increase in kidney transplants in relation to other transplants nationally. Instead, text in the subtitle and caption provide context for the visual and tell the viewer how the pattern they see can be explained by events not shown in the data.

4.12 MINIMIZING MISLEADING COMMUNICATION

While APs have the power to highlight features of an analysis in a way that helps to simplify or elucidate, visual manipulation is also one of the most common ways to mislead with data. As the individuals entrusted with guiding the application of analytics methods, APs should be cognizant of what misleading communication looks like in analytics. We'll cover a handful of representative examples, but in practice there are many more.

Visual manipulation of data is displaying data in a way that selectively leaves out part of the story, or leads the viewer to the wrong conclusion. The classic example of this is a bar chart with a modified y-axis. Consider the two data visualizations shown in Figure 4.4. On the left, the change may look like an impressive increase. However, if the axis started at zero, would it still be

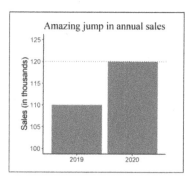 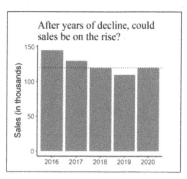

Figure 4.4 Example of a misleading chart.

perceived as such a drastic change? The graph on the right shows the change with corrected proportions and the context of the past five years leading to a different interpretation. The information that is provided to the viewer, as well as the information that is left out, has the potential to change the perception of data.

For certain values such as percentages and rating systems with set limits, it may be important to show a minimum value, often zero, as well as the maximum attainable for the variable. Displaying the full range of values gives the user an appropriate sense of scale by which to interpret results.

Sometimes the metric itself is the problem. For instance, consider a comparison of the number of doctorate graduates across the US states. California and Texas may stand out as having a high number of doctorate holders, but these states also have high populations overall. At a quick glance, one may simply take away that these states are more educated than others. However, with more consideration, the view may be left wondering, "out of how many?" Placing the information within its context, in this case relative to the population, provides a more meaningful result.

In more complex analysis, comparing groups that should be normalized is a common analysis mistake that can lead to misleading results. For instance, say you are using k-means clustering to find similar customer segments for an online retailer. If both customer age and approximate annual salary are inputs to the analysis, the salary, unscaled, could hold too much weight in model outcomes. In this distance-based method, the inputs should be scaled so that the features are considered equally by the model.

As analysts, we often need to provide approximations or are asked to package results as a single value. The single value or point estimate is in contrast to the reality, which may be a range of values, all with differing probabilities. The "flaw of averages" refers to the failings of this kind of oversimplification.

Disaster management provides an exaggerated example of this principle. When considering coastal flooding, city officials could not effectively make decisions off of the mean water level. Instead, they would need to know the range of possible water levels, the probability of each occurring, and the associated repercussions.

Most non-technical audiences can grasp the concepts of distributions and probabilities when communicated in familiar terms. Unfortunately, the mean is still an overused metric in practice. APs can advocate for more holistic, informed decision-making by showing the trade-offs of using a decision-making strategy guided by averages.

Units that are used for data is another possible point of confusion. Getting the audience to understand the units of each variable is crucial to maximize clarity. A method to increase understanding is to express units in the most familiar context for the audience. For example, fuel is measured by gallons when using many ground vehicles, pilots use pounds to measure the amount of fuel on aircraft, and barrels (42.5 gallons per barrel) is the unit of measure preferred onboard sea going ships. The same unit expressed three different ways depending on the context determined by your audience.

Awareness of these pitfalls in communicating analytics is the first line of defense for APs. The areas covered in this section highlight several major issues although there are many more that are industry or algorithm related. APs that identify and correct issues with miscommunication will greatly increase future project success.

4.13 SPREADING YOUR MESSAGE

The process of analytics is more than just discovery. It's also an opportunity for engagement and building trust with the public-which are all essential when it comes time to translate findings into policy or practice. When you share your work publicly in order to create awareness on critical issues, even if organizational issues arise from nondisclosure agreements and intellectual property; typically models/data can be generalized to allow for publication.

Dr. Laura Albert in 2022 challenged the analytics community to share their work in a Tweet.

Why public outreach? [5]

1. Improve scientific literacy

2. Shape conversations

3. Build trust with the public

4. Support for higher education & fundamental scientific research

5. Representation matters!

6. Support OR/MS ecosystem

You can get involved in public outreach in many ways. You can write for popular magazines or blogs, give talks at schools or community events, or participate in social media campaigns. No matter what approach you take, always remember to explain the science in terms that everyone can understand. Analytics is a complex field, but there are many ways to make it accessible to the public. By reaching out to the general population, you can help them see the value of our work and build a foundation of support for the future of science.

The only way to share analytics with the world is to be a part of that conversation. In the 2021 article in Phalanx, Harrison Schramm suggests three areas for action [32].

> **Message.** You have to have something to say. This something needs to be accurate, insightful, actionable, and (just a little bit) provocative. I cannot think of a single topic that we might comment on that would not consist of joining a conversation in progress. You need to know what has already been said in the "room." This is especially true if you plan to burn it all down.

> **Recipient.** You have to have someone to say your message to. In every engagement, I have an intended recipient in my mind. Frequently, the recipient is someone whom I have not met and may never meet. It does not matter if this is a person who does not exist, or even if they

are the director of an office that does not exist yet. Having a single recipient in mind tends to focus your effort and strengthens your thinking. At the end of the engagement, your hypothetical audience member should walk away and say "Wow, I never thought of [...] like [...]." Filling in these brackets is key.

Rapport. You have something to say, and you have someone to say it to. Great. Why are they going to believe you? This goes beyond credibility, which I had originally titled this section, and leads to rapport. What's the difference? Credibility means that the recipient should listen to you; rapport implies that we should have a conversation. Too frequently, we attempt to win credibility by assaulting our audiences with a barrage of awards, degrees, and papers. While this certainly does create an impression, it ignores their concerns, which are: Will this person listen to me as I describe my problems? And is this someone that I can talk to?

Think of analytics like the watermark that is helping shape our world. The more people who know about how this technology can be used for good, then sooner will come a time when they stop viewing it as some kind-of magical tool only found in high finance or healthcare.

4.14 COMMUNICATIONS EXAMPLE: KIDNEY EXCHANGE MODEL

Advocating for an analytics solution with broad impact can easily extend beyond workplace meetings and reports. For the researchers and the Alliance for Paired Donation (APD), sharing their work on NEAD chains took many forms. They not only needed to communicate the analytics, but also build rapport and find partners in the medical community. We'll look at how the APs on this project communicated their solution to other researchers in the field, physicians, and the world.

As donation chains grew in length, they also picked up media attention from sources including *People Magazine* and the *New York Times*. Although non-traditional for an analytics solution, this publicity helped the method gain acceptance in the public eye. Figure 4.5 shows a comparison of how a chain was represented

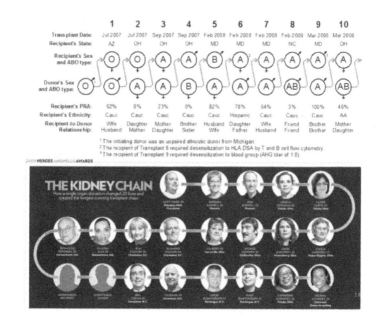

Figure 4.5 Diagram approach compared to using human faces. Roth, 2010.

in a medical journal vs how it was displayed in *People Magazine*. The visualizations were adapted to fit the interests of the intended audience.

EXERCISES

4.1 What unique challenges did the team of analysts working with the World Food Programme face?

4.2 How can the user design of analytics tools impact their adoption?

4.3 As an AP, what steps can you take to build rapport with a new stakeholder or customer?

4.4 Think about a recent analytics project in which you participated. How would you explain this project in five minutes to a fellow AP? What about a non-technical audience? What elements of the explanation would you alter for different audiences?

4.5 How could you use atomized knowledge to explain time series analysis? What subtopics could you break this concept down into?

4.6 Reflect on some of the most impactful data-based presentations you have seen. What set these apart and made them memorable?

FURTHER READING

LeFever, L. (2012). *The art of explanation: Making your ideas, products, and services easier to understand.* John Wiley & Sons.

Savage, S. L., & Markowitz, H. M. (2009). *The flaw of averages: Why we underestimate risk in the face of uncertainty.* John Wiley & Sons.

Silver, N. (2012). *The signal and the noise: Why so many predictions fail-but some don't.* Penguin.

Alda, A. (2018). *If I understood you, would I have this look on my face?: My adventures in the art and science of relating and communicating.* Random House Trade Paperbacks.

Krug, S. (2000). *Don't make me think!: a common sense approach to Web usability.* Pearson Education India.

Tufte. (2001). *The Visual Display of Quantitative Information.* Cheshire, Ct: Graphics Pr.

CHAPTER 5

Experience

> "Experience seems to be the only thing of any value that's widely distributed."
>
> William Feather

D R. AL ROTH, one of the key researchers in the development of NEAD chains, was an economist, with a research focus on market design. He recognized similarities between markets studied in theoretical economics and the observed scenario in kidney transplantation. The mathematics developed was inspired by economic research about trading indivisible goods.

Sometimes analytics methodology only goes so far, and APs need to supplement their experience with those that have subject matter expertise. After model development was complete, the analytics team started concentrating on implementation. To begin, that meant convincing one transplant doctor to use the results of the model. Since none of the analytics team had the medical expertise to perform the procedure, this necessitated identifying and convincing someone outside of the team.

Finding a physician that performed kidney transplants and could implement the new algorithm was challenging. After many attempts, Roth finally succeeded in convincing a friend and professor at Harvard Business School that was still practicing to use the model. This was an important step for two reasons. First, the model was implemented even at its simplest level. This first use could be documented and marketed to create more opportunities. The second reason was that the doctor that Roth convinced was now added to the team and his experience could help both to refine the model and he could be a proponent for future use. He provided actual clinical organ transplant experience.

Professional experiences unlock a deeper understanding of methodology by enabling analysts to forge connections between underlying mathematics and the real-world. In this chapter, we introduce the types of experience required for APs: analytics methodology experience, business application experience, and research experience. As many APs spend long stretches of their careers focused on a small number of analytics models, we present several ways to multiply experience through connections with other APs.

5.1   ORGANIZATIONAL EXPERIENCE

In technical professions such as analytics, the majority of research into the development and impact of experience focuses on the organizational perspective. From this viewpoint, the process of organizational learning consists of search, knowledge creation, knowledge retention, and knowledge transfer [8]. As the experience of the team is the sum of the experience of its members, these learning processes cannot be considered without overlap amongst individuals.

In organizational research, the purpose behind learning can be broken down into two categories, exploitation and exploration [21]. Exploitation refers to the repetition of tried and true processes to maximize efficiency and productivity. Exploration, on the other hand, refers to a process of experimentation with unknown outcomes. Organizations need a balance of both processes to be successful, but the emphasis placed on each that depends on the growth stage, risk tolerance, and goals of the team. The focus of the organization, whether on exploitation or exploration, transfers to the experience of individual contributors. It influences whether they will perform repetitive, high-efficiency tasks or spend their time exploring new concepts. The organizational strategy determines the need for individual focus on specialization or generalization. While specialized experience has been found to have the most impact for individual contributors, diverse experience has shown to be most crucial for overall team performance [14].

5.2 TYPES OF EXPERIENCE

The three types of analytics experience are: analytics methodology experience, business application experience, and research experience. The importance of the three types varies by an APs intended area of work, but a basic level of each empowers a balanced approach to analytics. An AP with only analytics methodology experience will be able to provide an accurate reading of any given analytical work product but may not understand the context in which the analytics are being applied. A successful AP with business application experience understands the data sources, business rules, or assumptions associated with a given analytics problem. They also understand the wider context within which the methods are being applied. Finally, an AP with research experience can efficiently invest their time finding new ideas and combining those concepts in interesting ways to generate innovative solutions to analytical and business challenges.

5.2.1 Analytics Methodology Experience

Experience with methodology is the foundation of a career in analytics and the starting point for most APs. Analytics is a field that requires broad technical skills, including, but not limited to quantitative analysis, statistical modeling, and visual design. Not only do you need to know how to use a particular method or technique from start-to-finish, but you also need to understand its advantages and limitations as well as accompanying statistical tests. This type of experience extends to familiarity with certain statistical, programming, and visualization software that support various methods.

Analytics methodology experience is also important in analytics consulting when building analytics systems or when developing analytics standards. Standards require a deep understanding of analytics methods and the industry's analytics taxonomy.

5.2.2 Business Application Experience

Business application experience is what is most often referred to as "real-world" experience, when an AP takes textbook analytics methodology and applies it to a problem with all of the complexities of the real-world. An AP with business application experience excels at demonstrating how insights can be applied within an organization, and at interacting with stakeholders to understand business needs and requirements. This type of experience is also particularly helpful to analytics consultants when discussing analytics use cases and understanding analytics demands.

In order to apply analytics, the ability to "speak the language" of the team, organization, and industry in which you're operating is indispensable. This language can be learned through domain knowledge or specialization in an area of application. For instance, one AP working within a hospital may need domain knowledge of patient care to improve outcomes, while another may specialize in anomaly detection in medical diagnostic scans. The language required to communicate with co-workers, such as hospital administrators, doctors, or clinical researchers is very particular. While these two APs could very well come from the same background, they could not switch roles for a day as the need for specialized domain knowledge is too high.

5.2.3 Research Experience

What if an analytics or business problem requires a methodology that is outside the area of experience of the AP? APs must be able to build upon what they have used in the past and explore new approaches. The most efficient and reliable way to do this is often through reviewing peer reviewed journals and articles.

In a time where the internet is brimming with articles claiming to have all of your analytics needs covered, it's important that you become familiarized and comfortable reviewing peer reviewed literature. These studies are usually more reliable than non-reviewed sources because they go through a rigorous process before being published-and their publication status gives them credibility when using these methods for data analysis or reporting purposes in future projects. Reasons why this is important are:

First, there is a firm foundation to reference and cite when using the new method. Otherwise, there could be the perception that the method was created independently. Second, the articles always provide information on the strengths and weaknesses of methods. Knowing both sides will help the AP access and communicate the risks of using these methods.

Many journals are open access, but for those that require membership, access to peer reviewed research is available through several sources:

- Academic libraries – Access is provided to current faculty and students. Alumni may also be granted access to library resources.

- Professional Societies – Societies may provide access to journals and articles for free, included in the membership dues, or as an additional fee.

- Direct Access – Journals provided for subscription directly through a website. Most journals offer an online and print version.

Innovative ways of approaching problems rarely occur in a vacuum. Typically there are several groups working to advance the science and technical fields, which should be considered when looking at new methods or approaches for problem solving. This gives you access not just one perspective but multiple points-of-view so that there is no blind spot on how these innovations might save time or effort down future implementation paths.

5.2.4 Putting It Together

Combining the three types of analytics experience enables the development and deployment of innovative approaches. Consider for example the process of selecting an approach for a new project. An experienced AP will need not only methodology and research experience to know their options, but also subject matter expertise to fully understand the nuances of the underlying problem.

An AP with analytics experience will have two layers of analytics knowledge: analytics methodology and analytics applications. This allows them to interact on equal footing with analytics specialists, advanced APs, business analysts, data scientists, and non-analytical decision-makers.

5.3 SPECIALISTS VERSUS GENERALISTS

One of the most common questions for early to mid-career APs is whether to specialize or generalize their skills. Is it better to have deep experience in one area of analytics or broad experience across a variety of analytics applications?

As more and more companies are employing advanced analytics, the need for specialists is on this rise. In analytics, specialization is generally more technical and academic in nature, with APs placing deep focus on one or a small set of projects. Specialists are highly valued and sought after, often more commonly found at organizations with an extensive analytics infrastructure. Of course narrowing your area of expertise also narrows the field of opportunities at which you can apply your specialty.

APs at small to mid-size organizations may operate on a small team or even be the only analytical professional in the organization. Flexibility and the ability to support a wide variety of analytical tasks is essential in this role as APs may need to juggle many projects at one time. Generalists have the ability to manage the complexity of real-world problems, find solutions, and make connections that specialists may not. With the changing nature of analytics work, generalists will see the importance of being able to learn quickly and adapt to different working environments.

In reality, the vast majority of analytical roles are a blend of specialist and generalist skills. For instance, you may be the go-to AP in your organization for Natural Language Processing tasks, but still work on a wide spread of analytics problems and applications in your day-to-day. Given a long career in analytics, you will have roles that are more specialist in nature and others that are more generalist.

5.4 LEARNING THROUGH FAILURE

Failure is an event or occurrence that prevents success. It can occur at any stage of a process, from the initial planning to the final evaluation. APs must have a clear picture of what a successful solution looks like when learning from failure.

Typically, the success of an analytics project is defined by the use of the model results. This may be a one-time recommendation to inform a decision or the implementation of a model that makes a periodic recommendation. If this is the goal, then anything less than the use of the model output for decision support has failed in some way.

Some of the most impactful learning experiences take place when things go wrong. In medicine, Morbidity and Mortality (M&M) meetings provide healthcare professionals with a forum to openly discuss errors and adverse outcomes. During a meeting, a resident presents a case to the group, sharing the patient history, what actions were taken, and the eventual adverse outcome. The audience then has an opportunity to ask questions and share stories from similar cases. The focus of these sessions is on improving systems, not on blaming individuals. Often, the group can agree whether or not the adverse outcome was preventable and create a plan to ensure it doesn't happen again. Studies have shown that

M&M meetings have a meaningful and tangible impact on patient outcomes [34]. By discussing failures and sharing the scenario that led to them, collective learning can take place.

Fortunately, in analytics projects, the stakes are not usually as high as life or death. However, the frequency of failure in analytics is shockingly high. In 2019, Gartner predicted that only 20 percent of analytics insights would deliver business outcomes over the next three years [3]. With analytics projects, it can be difficult to define and measure failure as most failed projects do not "crash and burn." Instead, a model may simply fail to have an impact with its influence fizzling out over time. For this reason it is important to clearly define both short-term and long-term project success. By monitoring and measuring these over time, APs can better understand what is effective in their organization and pinpoint weaknesses.

Over the course of a career, some APs may only focus on a handful of large scale projects. Further, as individuals change roles and organizations, they often do not get to see the full lifespan of a project they supported. With this narrow exposure to projects, individuals are limited in truly understanding the success of analytics and the possibilities for shortcomings in implementation. This challenge is why sharing project stories is so essential for the profession: openly communicating about project outcomes within an organization or even within the broader community allows for a multiplicative effect on learning. In practice, "wins" are overwhelmingly the focus of conferences and publications. In depth assessments of failed projects rarely happen and these stories are rarely shared outside of an organization. While it is undesirable for a team or department to share their own missteps, it could be incredibly valuable for others to learn from them.

Fortunately, the tide is changing. Acceptance of failure as a natural part of doing business and an important tool for continuous improvement has led to more open sharing of experiences. After all, nothing great was achieved without the risk of failure.

5.5 STARTING OFF: CREATING WAYS TO GAIN EXPERIENCE

Experience is a tricky thing: it can seem impossible to gain it when you have none and feel trivial when you have a lot. Limited real-world experience poses a challenge for not only new graduates, but also those transferring into analytics from another field. Several ways exist to bolster textbook analytics knowledge with real-world application without formally working in that area.

Creating or finding your own analytics experiences outside of work. A wealth of datasets amenable to complex analysis and data science are publicly available online through data portals, government websites, and learning platforms. APs can build their own analytics experiences through portfolio development. Assembling a portfolio of personal projects allows you to explore data and topics of interest while gaining experience in unfamiliar methods. APs can also leverage the analytics community by engaging with local or global professional groups, meetups, and societies. Volunteering for these kinds of groups is a great way to learn more about the profession while making valuable connections. You can also find exciting pro bono opportunities through which you can gain real-world experience while making a difference for a cause that has limited resources.

Spend time learning from others. This idea extends beyond the technical piece of analytics to seeking out stories about projects and the day-to-day for career professionals. Studies have shown the effectiveness of experiential knowledge transfer within teams [9]. Podcasts, blogs, and community events are valuable tools for making these connections.

Mentorship. This is another extremely valuable way to not only gain knowledge, but also clarify professional goals. Mentors can help by sharing challenges in the profession and how they overcame them. For those with more experience, mentoring a newer AP is also an important learning opportunity for the mentor to approach problems from a fresh perspective.

APs at any stage of their career can benefit from these varied experiences. These are especially useful to aid in the transition from one area of analytics to another.

What's the best way to get analytics experience? Practice is the best teacher. As with any skill, APs need to practice their analytics applications and analytics methods knowledge extensively before they are ready to apply analytics in an analytical way.

At its core, analytics is the combination of mathematics with information. Expertise in analytical methods requires an entirely different skill set from expertise in subject matter context. Successful APs require some combination of these two elements: business application experience and analytics methodology experience. The importance of the elements varies widely. Some very successful APs operate in a research environment, surrounded by like-minded individuals, advancing analytical methods. For most, the need for deep knowledge in the area in which they are applying analytics is essential.

Many roles and particular tasks require a high degree of both aspects of experience. A perfect example of this is cleaning up a messy dataset. This task demands both the technical know-how for methods such as interpolations and the subject matter knowledge to identify and properly handle anomalies in the data.

When learning a new skill, do what medical professionals do if you want to explain the element quickly and succinctly. They have a saying, "see one, do one, teach one." First, you observe someone else performing the skill. Any YouTube, online website, or real-life exposure will do. Second, you then perform the task yourself several times. Lastly, you teach the skill to another person. The person doesn't even have to be an analytics professional. A friend or spouse will do nicely. Adding "write one" to this sequence will further help reinforce the long-term knowledge retention. Writing could occur in a personal journal, professional journal, or an organization's blog.

5.6 ANALYTICS SELF CARE

One of the great things about analytics is that it is a dynamic profession. Technology advances, tools change, and the scope of the traditional analyst role is ever-evolving. This can be a challenge, but it is also what makes the profession so interesting.

Many options exist to continue learning in analytics, enabling you to gain new knowledge when you feel you've plateaued. This could involve taking courses, reading books or articles, attending conferences, or listening to podcasts.

Once you have made the effort to learn a new topic, consider teaching what you know to your team or colleagues. This can help them to improve their skills and it also helps to cement the knowledge in your own mind. You can also contribute to the community through research, collaborating on open source projects, participating in competitions, or involvement in professional societies.

All of these activities help you to continue learning and keep your skills sharp. By combining your non-analytical interests with analytics, you can also come up with innovative ways to solve day-to-day challenges. Lifelong learning is essential for any successful AP.

Examples of tools and datasets to expand your analytics skill set:

- Organizations that support the use of analytics to achieve social good

    - Data 4 Good – Local chapters of Data 4 Good can be found across the world, supporting data-based projects for a better world.

- Competitions

    - Hackathons – Initially created by the computer science community, hackathons are a great way to meet people and test your skills in data science and analysis.

    - Online competitions – Websites like Kaggle present opportunities to solve data science challenges independently or with a team.

    - Tidy Tuesdays – For users of the statistical programming language R, Tidy Tuesday is a weekly activity to practice wrangling, visualizing, and modeling new datasets. Participants can post their output on social media for feedback from the community.

- Open Source Datasets

    - Open Data – Local datasets covering topics such as a city or region's population, environment, or transportation network provide a way to expand your skills while learning about a location. Open data sites often have spatial datasets available.

    - University of California Irvine Machine Learning Repository – This collection is a well-known source for machine learning datasets.

    - Data.gov – Covering topic areas such as climate, energy, and local government, this website is home to the US government's open datasets.

Here are some suggestions to exercise leadership and help further your career development in analytics:

- Create a company-approved project that will force you to learn new skills and introduce you to new people within your company

- Take on leadership positions in the hobbies and outside organizations that interest you

- Join your local alumni club and spend time with people who are doing the jobs you'd like to be doing

- Enroll in a class at a community college on a subject that relates to either the job you're doing now or a job you see yourself doing in the future

EXERCISES

5.1 How did the three types of experience, methodology, business, and research, come into play for the kidney exchange researchers?

5.2 Do you consider yourself to be a specialist or generalist in analytics? What are the advantages and disadvantages of each approach to experience?

5.3 Have you participated in an analytics project that was not successful or deviated from the initial plan? What lessons did you take away from the experience?

5.4 What ways could you create new experiences in analytics?

FURTHER READING

Epstein, D. (2021). *Range: Why generalists triumph in a specialized world.* Penguin.

Convince Them

W HAT IF we could track and predict outbreaks of a virus before cases are confirmed by in-person physician visits? This foresight would allow healthcare providers to prepare resources and reduce the overall impact of a virus on the community. In 2008, Google attempted to do just that for the seasonal influenza with the search engine-based tool, Google Flu Trends. The idea behind the tool was relatively simple: utilize trending search queries to predict regional influenza-like illness physician visits within the United States. For example, if someone enters a query such as "do I have the flu?" or "flu symptoms," there is a strong likelihood this searcher is experiencing flu-like indicators. By using the wealth of search query data available, Google could crowdsource flu outbreak detection.

The Flu Trends model was based on recognizing and utilizing search terms that displayed similar patterns to flu cases in the past to predict future flu instances. As searches for the flu increase, we can expect cases of the flu to increase as well. Of course developing the actual model was not this simple. The Flu Trends developers were aware of potential issues that may plague their model and accounted for them as documented in their 2008 paper on the methodology [16]. While the top predictive search terms the model identified were largely precisely related to the flu, seasonal search trends also found their way into some of the top matches. For instance, the search query "high school basketball" was recognized within the top 100 best indicators and removed.

While Google Flu Trends was at first praised as a shining example of the impact potential of big data, it soon after became known as a warning for the risks to future efforts. What went wrong with the long-term implementation of the Google Flu Trends model?

After a few successful prediction years, the approach faltered, underestimating a seasonal outbreak in 2012 and overestimating an outbreak in 2013. The prediction missed the mark due to an unexpected change in the data. Search topics changes dramatically with time, capturing trends, seasons, and even changes in language. A critical limitation of search data is that it only reflects what people are searching for, not their reasoning behind the search. For instance, if an individual hears of a local viral outbreak on the news, they may search for "flu vaccine," "flu outbreak," or "flu symptoms" to educate themselves, not necessarily because they are experiencing influenza-like symptoms. As such, highly publicized outbreaks could be overhyped and overestimated by this system.

The model lost public trust as a result of its inaccurate predictions and was sunsetted by Google in 2015. While many argue that this task is still amenable to analytics and more suited to approaches developed in the late 2010s, it remains a cautionary tale for APs.

6.1 DELIVERING THE COMPELLING ANALYTICS MESSAGE

Convincing them is the end goal of the AP. This statement doesn't refer to always being right or using analytics to sway decision-makers one way or another. Convincing them is about showing the worth of analytics, implementing models that align with business needs, and getting others onboard for the change that a data-centric approach can initiate. In this stage, the elements of trust, experience, communication, and analytics come together, all tailored to the audience.

If the analytics falters in implementation, as with Google Flu, it can negatively impact the trust between the customer and the AP. The same is true of a very successful model. If a customer sees immediate business value derived from a model in production, trust will increase. In this chapter, we present three dimensions of model success, explore the interactions between elements, and share a fictional story of the elements in action.

6.2 DIMENSIONS OF MODEL SUCCESS

A challenge in measuring the success of an analytics solution lies in unclear goals for its lifespan. As APs, we want our models to be "useful," but what does that really mean and how can it be measured? Anderson and Hoffman introduced three dimensions of Operations Research which can be applied more broadly to analytics [6]:

> **Installation.** A model that has been installed is one that is operating with real-world data and is available in an environment or existing workflow to support decision-making. APs today may call this process "deployment."

> **Implementation.** Implementation takes place when the model is actually put to use to better describe, understand, or predict an aspect of the business problem. It has a tangible impact.

> **Integration.** Integration refers to the cultural shift that occurs as users of analytics models begin to recognize their value, promote data-backed decision-making, and seek out future applications of similar approaches across multiple domains.

One would expect these three dimensions to build progressively on one another. In practice, this is not always the case. A model can be successfully installed, but fail to have a broader impact on the organizational attitude toward analytics. Alternatively, decisions may be made while influenced by a model that is not fully installed. A state of integration is rarely achieved without several successful implementations of models.

The three dimensions of success may also be understood uniquely by different "Convince Them" audiences. Surprisingly, there are instances where the target audience may be more concerned about a non-analytical aspect of the project or keyed in on a seemingly trivial side effect. For instance, when implementing a refueling ship model for the US Navy, stakeholders were more concerned about the automation of the daily 10-page scheduling message than the actual model. The manually developed message was created by hand and had to pass through multiple reviews before dissemination to the ships. Automation of this message saved several hours a day and made life easier for the users [12].

Asking the stakeholders and users questions such as "What would it take for you to use the model or trust the analysis?" are insightful and may lead to requirements that are not directly related to the analytical model.

6.3   CREATING AND SUSTAINING THE CHANGE

Believe it or not every target audience does not need to be a zealot for your analytics solution. Each person working with an organization possesses a commitment level at any given time. The possible levels of commitment are:

- **Unaware.** This is when a person is ignorant of the solution.

- **Aware.** In this state, the person has some knowledge of the possible solution. Perhaps they have heard the name of the solution or are familiar with several of the team members. They may even know what the objective of the solution is.

- **Understand.** A person has some background knowledge on the history and objectives, and is familiar with the current status of the analytics project. They can explain at a basic level how the solution works and what are the model inputs and outputs.

- **Adoption.** An individual is willing to work with and implement the vision. They support the effort fully.

- **Committed.** An advocate for the solution who helps create innovative ways to use and improve it. These individuals not only get it but are willing to make the analytics solution their own. They will make the most suggestions and be advocates in their own organizations for the solution.

Influencing people from the level of unaware to committed can take a tremendous amount of time and effort. Moving someone from unaware to aware may only require a few emails. Moving them further to understand can take emails, multiple presentations, and possibly training sessions. To get to the adoption level even goes further with potentially one-on-one discussions and training. To obtain the highest level takes the most amount of time. At this level an individual is so familiar with the analytics solution that they are creating enhancements. This may require them to have used the solution for a long period of time to build trust. Many types of personal interactions with these individuals both for training and for feedback are possible. At this level, they must feel like they own the solution.

Since no project has an infinite amount of time, an AP must determine which decision-makers and stakeholders need to be at what levels. Typically many successful projects have at least one committed individual at the leadership level, middle management, and at the worker level. In larger organizations, there may be the requirement to have at least one committed individual in each division or department. Again, the time and resources it takes to develop this level requires optimizing this decision.

6.4 HOW THE ELEMENTS INTERACT

Over the course of an individual AP's career, there are trade-offs between the importance of each of the variables discussed in this book. In some projects, the analytics may be straightforward, while crafting the communication surrounding it may require much more effort. In others, the AP may already have the trust of stakeholders and could need to place the majority of their focus on a novel analytical approach. In this section, we explore how each of the elements discussed in this book support or detract from the development of the other elements.

6.4.1 Analytics - Trust

Analytics is a valuable tool for progress monitoring, prediction, and objective decision-making. But analytics can also be used to make decisions without taking human sentiments into account. This absence of direct human oversight may lead to distrust forming between parties involved in a business transaction or project. It is important to remember that humans are integrated in the analytics development, approval, and implementation process so human opinions are always a factor.

While lines of communication remain open, strong trust in an AP-customer relationship often takes the form of more hands-off involvement in the development of a model. This confidence in capability offers the AP more freedom in the analytical approach applied. While having trust does not necessarily mean that the customer will be less involved as this is highly dependent on working style, it usually means they will be less invested in influencing the minute details of a model. When trust is absent, managing overextended customer oversight can be time-consuming and impact the final quality of a solution.

Data access is a crucial topic for trust and analytics. Not every customer has the desire or ability to communicate their data in a standard format. For many organizations, data is one of if not the most valuable asset. In cases where it is difficult for a modeler to gain access to data, they must be prepared to adjust the type of data analysis conducted. Building a relational data model using "unknown" variables is one option, but there are other options that may be applicable to the case. In some cases, an AP will still find themselves in a position where it is difficult or impossible to acquire data from a customer, and at this point it is time to part ways. The last thing an AP wants is to put time and effort into a promising modeling process, only to fail at implementation due to the inability to acquire the right data.

6.4.2 Analytics - Communication

The relationship between analytics and communication is likely the most discussed in analytics textbooks. Proper communication increases the "stickiness" of an analytics solution, but that precise level and mode of communication can be difficult to find.

When it comes to communication, keep in mind that all communications will most likely be one-sided at the beginning of the relationship. The audience's goal is to understand what is possible with analytics and how that fits into their organization. It is up to the modeler to position each solution within the context of their customer's business to ensure that expectations are set appropriately. Even if a solution has been implemented in the same industry before, it is important not to take for granted that this will be an easy sell. Each business is unique and additional analysis must be conducted in order to determine how much effort should go into the modeling process. Presenting a model without proper preparation can cause a customer to lose interest and a modeler can quickly find themselves losing a customer.

The need to clearly communicate can also impact the selection and development of an analytics method. For instance, if you know a customer requires full traceability and comprehension of every aspect of a modeling approach, this may steer the project away from the use of black box methods, even if the data is amenable to them.

Each project requires a unique blend of the above elements, and these requirements may change over time. None of the elements can be considered without the influences of the others.

6.4.3   Analytics - Experience

Experience is a key factor when it comes to analytics. The practice of analytics, successful or not, informs the next time an AP approaches a problem. It's important to have experience with the data with which you are working. This familiarity will help you to better understand the problem and make better decisions. With more expertise, an AP can quickly and accurately diagnose an analytics problem. The more experience you have with your data, the faster you will be able to achieve accurate results. The experienced AP knows common pitfalls with an approach or with the data surrounding a familiar problem. However, practiced APs may be reluctant to try new solutions, when they know the result of a tried and true method. Similarly, they may overlook a nuance in a familiar problem that a newer AP may question.

If you do not have experience with your data, context, or business question, the analytical solution could be negatively impacted. Not only could the solution be incorrect, but it can take longer because you must test out different possibilities. In this case, it is important for an AP to rely on the expertise of others in the field by consulting senior APs or referencing literature. However, many APs think that they need to know everything about every field. Although this is important for certain fields, it is less important for others. If an AP knows everything there is to know about sales, it might be difficult to apply those analytical skills to other areas. The more experience you have with particular fields, the simpler it becomes to find patterns and relationships within those fields.

6.4.4   Trust - Experience

Experience is an obvious differentiator. Generally speaking, individuals with more experience garner more trust from their audience. Someone who has successfully solved a problem twenty times before provides more assurance than a fresh graduate who has never worked with real-world data. In reality, individuals' experiences are varied and cannot be compared by years alone. Say a customer must select one of two modelers. The first has experience in many industries and demonstrates superior skills, while the other comes with industry specific experience and successfully delivers solutions to the same verticals over time. The customer must choose wisely knowing that each will require trust in order to continue building a relationship with their chosen AP.

One's experience is only relevant when communicated effectively. Otherwise, it may be lost to the audience. Similarly, if an AP with limited experience is able to communicate the impact of an analysis, they may build swift trust with a new audience. While experience impacts trust, the capability to build trust one also impacts experience. If you have trust, you open doors to opportunities to do more analytics or to be more creative with your analytics approaches. The skeptics are often the ones that are stuck in their analytics processes but when they start adding trust into their analytics workflows it allows them to become more flexible and adaptable with analytics tools allowing for easier implementation of analytics processes.

6.4.5 Trust - Communication

Rapport is the key for tying together trust and communication. If you want to build a strong working relationship with someone else, then you need to learn how to communicate well first. With rapport, someone can feel your honesty and sincerity as well as see it in your body language. However, if lines of communication and underlying trust is severely damaged, then the relationship will be negatively impacted. Communication without trust is just an illusion.

Trust can also determine the manner and frequency of communication. A positive relationship built on trust between an AP and a customer will lead to more open communication. We are more likely to share minor concerns and compliments with someone we trust. These small pieces of open feedback can lead to significant changes in overall alignment with customer preferences and satisfaction.

6.4.6 Communication - Experience

In order to be a good communicator, it is important to be able to understand the audience and their needs. In cases where you are not familiar with your intended audience, you will need experience in order to do so. The more time an individual has spent communicating within a certain context or group, the better they are at understanding what is expected of them as well as how best to approach the situation. While experience can provide insight for understanding who your target audience may be, there are also people who have never had any prior exposure that still possess excellent communication skills. For these individuals, practice might help them improve even further or reach new audiences by exposing themselves to different contexts and perspectives over time. Like many other things in life though, the more effort and energy put forth in order to better communicate with others, the more readily someone is able to do it.

6.5 DATA AND DENIM

Now that we understand the concepts, let's see how they work in a fictional story. We start our story with Carla, a Principal Data Scientist working at Company ABC that manufactures denim apparel. She leads a small team of data analysts and data scientists who work closely with process experts to improve throughput, reduce costs, and keep up with the latest advancements in sustainable apparel manufacturing. Carla's team is relatively new and several executives are unconvinced of the advantage of using data-backed decision making in their industry. ABC just opened a new factory in Los Angeles and their presence has been met with some disapproval from the community.

6.5.1 Understanding the Business Problem

As Carla walks into the ABC offices, she sees her coworkers were gathered around a computer. Over the weekend, a local newspaper published a feature article about ABC's environmental impact on the surrounding ecosystem and community. Local residents have been concerned about the high resource usage and harmful outputs of their production process. Carla goes to her desk and checks her email to find a request from the company leadership. While their marketing and legal teams craft an initial response to the article, engineering needs to make significant alterations to their manufacturing process to align with environmental regulations. Her team has been tasked with determining what changes to the production process they could implement in order to reduce negative environmental impacts without hurting profits. The data team has three months to advise leadership on process changes before they meet with environmental regulators.

During the afternoon, Carla sets up a tour of the new manufacturing facility for her team. As they walk through the factory, they note the inputs and outputs of each process, paying particular attention to the resource usage. The full process requires significant energy and water to create the end product, and harmful chemicals used in the dyeing process endanger local waterways. Carla takes note of where data collection is already in place and where it would need to be added to support analysis. After the tour they meet with a group of process engineers that share several options for improving the environmental footprint of their operations:

- A new filtration system on the market that would allow them to recycle 98 percent of the water used in the production process is found. Unfortunately, this system is expensive to install and would take up the majority of the available funds.

- Synthetic indigo dye is particularly dangerous and is being released into local waterways. They could implement process changes to increase the amount of dye captured.

They acknowledge there could be other areas in the process with room for improvement: e.g., reducing energy usage, but these will require a study to better understand and assess.

6.5.2   Drafting an Analytical Approach

Carla sits down with her team to plan out their approach for the next few months. She explains that ABC leadership wants to better understand the drivers behind its environmental impact and have requested recommendations on how to modify the production process. To start, they break the business problem into two analytics problems:

- What is the trade-off between implementation cost and environmental impact of each proposed process change?

- What areas of the process are the most costly?

Their analysis begins as a series of smaller projects through which her team attempts to collect data and model small parts of the manufacturing process. In the past, environmental data was collected on an adhoc basis, not monitored and managed like other aspects of the production process. An analyst on her team drafts a plan for the installation of sensors to monitor water and energy usage. Daily data can be stored in an existing database, but the set up will require resources from the IT team. She presents the plan to the head of the IT department and they put in a request for funding.

One analyst performs a cost-benefit of the options and finds that the new filtration system is too expensive to implement at this time. They find that there is not enough information about the other process changes and recommend the team assess the full process. In the meantime, the new data collection system is approved and put into action.

6.5.3 Convincing the Team

With all of the supporting data now accessible, Carla starts by mapping out the energy used at each step of the process. She notes that the drying of the denim is one of the most energy intensive and variable steps of the process, accounting for approximately 20 percent of the overall energy usage. With this knowledge, her team developed a short five question survey that took 5-10 minutes to complete. The survey was emailed to the process engineering team. The results of the survey revealed that ten of the engineers had prior experience with alternate drying processes. Carla scheduled a focus group with these ten process engineers. One of the engineers worked in a factory overseas that air dried their denim in order to reduce this requirement. The climate in Los Angeles would be conducive to this process change, but is it feasible?

As the project grows, Carla's team must convince her that they are using the appropriate analytical methodologies. A data scientist on her team suggests that they use a black box approach to predict energy savings trained on data from a small scale test, but Carla has concerns about the interpretability of the solution. How will they explain this approach to the members of leadership? Will the model answer the right business question?

Carla attended a professional conference and saw a brief where the research used a discrete event simulation to model a complex healthcare process. She accessed the research paper that was presented and several other referenced papers. Interested to learn more about the methodology, Carla volunteered to assist in a Data-4-Good event in her hometown. The event occurred over the weekend and consisted of several teams of volunteers using analytics to solve problems related to public issues. The problem that piqued Carla's interest was a city bus station passenger loading problem. Carla volunteered to participate on this team. She built a simple discrete event simulation that provided insight into modifications that the city could make to make the passenger loading more efficient.

After this event, Carla wrote about this experience on her blog. She solicited feedback on the strengths and weaknesses of this approach and for others to share examples of their use of this method. Combining the knowledge through her research, the experience she had gained during the event, and the feedback she had received after posting her blog post, Carla was ready to propose this approach to her team.

After presenting and discussing the analytical approach with her team, Carla convinced them that discrete event simulation was worth moving forward in developing. Back at the plant with the process engineering team, air drying was showing promise, but hadn't yet been implemented at this scale and a simulation would allow them to better understand its impact before making a financial commitment. From her experience working with fashion executives in the past, Carla knows the results of the simulation will be easily understood by the stakeholders at ABC.

Once the solution is developed, Carla must convince the director of analytics that her team is using the data and methodologies appropriately. She shares several examples from conferences and research of similar approaches currently in operation, explaining the similarities and differences of each in comparison with their problem. Next, the director asks to be walked through a few key areas of implementation in the code for this solution. Carla is glad to find that her team has properly documented and commented their work, making this meeting easier. The director approves the selected approach and sets up a meeting with company leadership to share their recommendations.

6.5.4   Convincing Company Leadership

ABC company leadership is not math-oriented and skeptical about the value of the new changes. Carla is familiar with the sometimes clashing priorities of the key decision-makers. While the head of operations will be most interested in the financial and feasibility aspects of the changes, the marketing director is particularly interested in the environmental aspects. To conceptualize the trade-offs between these differing priorities, Carla's team creates a simple application that allows users to test out different process options:

- Where to position the racks to take advantage of high-heat areas in the factory,

- Whether or not to use fans,

- Whether the denim moves or not,

- And whether to finish drying in a conventional machine, and if so, at what percent dryness?

For each configuration, the model outputs the installation cost, energy reduction, and time to dry, which will impact their throughput.

Together with teams from legal, operations, and engineering, Carla's team is briefing company leadership to propose the plan for moving forward with process changes. This project is presented to the company leadership team over three monthly presentations. The lead for the operations team kicks off each brief by providing an overview of the project and status update on the schedule. Carla is the next presenter. Her brief is structured into three sections.

Carla begins her presentation by introducing herself, her role on the project, and her future impact with her team's analysis. She introduces the analytical approach, connecting the analytical problem with the business problem. This includes one slide in each brief that introduces a different mathematical concept behind the discrete event simulation that was used to model the process. During the first monthly brief, Carla introduces the triangular distribution and explains how they have used it to provide a rough estimate in the absence of observed data. In the second monthly presentation, she introduces the concept of assigning the appropriate distribution to each process, and during the third monthly brief, she introduces the concept of simulation output having a range instead of a point estimate.

The second section of the brief discusses the input and output data. This includes what inputs were controllable (dials) and inputs that were constant. She introduces the application their team built to explore the process options and they test several configurations together to build trust in the solution. The third section of the brief focused on the results and findings.

Carla's presentation was scheduled for 30 minutes. She practiced delivery of the slide deck so that it could be comfortably delivered in 20 minutes, allowing 10 minutes for questions and discussion during the brief. By the end of the three months, they decide on a configuration of process options and start planning for installation of the new drying racks.

Carla schedules multiple focus group sessions to get feedback on the solution. The first focus group is scheduled with the operations team. The meeting provides several areas of concern over several assumptions on facility ventilation. The second focus group discussion is with the accounting team to validate cost calculations and assumptions. The third focus group is with the legal team. Carla is interested in incorporating as many legal concerns as possible into the model.

6.5.5   Convincing Environmental Regulators

After the company leadership is convinced and the company implements the methodology for determining the reduction in carbon footprint, the company is called upon by regulators to justify their methods. At this point, Carla and her team must collaborate with environmental experts to develop the message for regulators.

Beyond filling out the required paperwork, her team needs to support development of an environmental impact report. In the section she is authoring, Carla needs to show the expected improvement from their previous state. The regulators require organized, labeled data and code to assess their solution. Carla prepares the required materials and works with the rest of the team to craft their message.

6.5.6   Elements Required

Each time Carla and her team needed to convince an audience, from fellow APs to company leadership and environmental regulators, they needed a different mix of the elements outlined in this book. Figure 6.1 shows a breakdown of the elements at play in each of these scenarios. While the beginning of the project and meeting with the director of analytics were focused on the

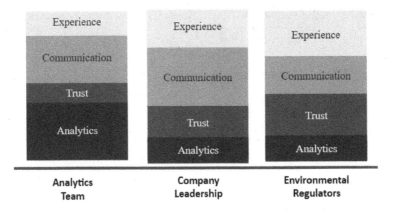

| Analytics Team | Company Leadership | Environmental Regulators |

Figure 6.1 Convince them elements used with each audience.

technical aspects of their work, the elements of communication and trust gained importance later on in the project.

After submission of the report, Carla gets a call from the director of analytics: ABC leadership was so impressed with the environmental and cost savings of the air-drying initiative that they would like her to visit another factory and perform a similar assessment. A few executives have even asked for access the application that they shared in order to test out some additional scenarios. Carla puts down the phone and congratulates her team. She's optimistic that the data culture at ABC is changing and that their work is having a real impact.

EXERCISES

6.1 Do you think Google Flu Trends could be successful today? How could the approach be adapted to be less sensitive to the nature of search data?

6.2 Describe how two element interactions played a role in the success of Carla's project.

6.3 What levels of commitment did the various stakeholders at Carla's company have to be at to implement the model?

6.4 Which dimension(s), installation, implementation, or integration, did Carla's modeling effort reach in beginning and at the end of the project?

FURTHER READING

Heath, Chip Heath and Dan. 2010. *Switch: How to Change Things When Change Is Hard.* 1st ed. New York: Broadway Books, 2010.

CHAPTER 7

Conclusion

"We can only see a short distance ahead, but we can see plenty
there that needs to be done."

Alan Turing
Computing machinery and intelligence

THE FIELD GUIDE to Compelling Analytics offers a rec-
onciliation between the promise and reality of analytics. In
the application of analytics, things don't always go as planned and
often the technical aspects of projects turn out to be the most
straightforward.

7.1 WHY THE HUMAN ELEMENT MATTERS

As long as humans are involved with either developing or implementing analytics, factors such as trust, communication, and experience are always relevant. Human interaction is a process that has
been refined over thousands of years. Over that time frame human
accomplishments have changed the world. When it comes to analytics, why would anyone think that the human element wouldn't
matter? Understanding that the practice of analytics remains a
human to human endeavor is the key to an AP's success in the
field.

What should a brand new AP focus their efforts on when they
are just starting out? All of the elements are important, however,

while trust and experience are attributes that an AP builds and maintains over the course of their career, communication and analytics are action-based. Communication and analytics are the most powerful for an inexperienced AP to succeed in convincing an audience of implementing their analytics solution. New APs can increase their competency in these two areas at a faster rate.

7.2 THE HUMAN ADVANTAGE OVER AUTOMATION

Without question, automation has had a lasting and positive impact on analytics. By automating certain processes, APs can do more in less time. Establishing automated workflows allows a more hands-off approach creating subtasks that can be monitored by others. Automation is powerful for data and modeling pipelines, anomaly detection, and even exploratory analysis. It can be implemented to provide coverage for one step of the analytics cycle, or with modern dashboarding tools, can undertake full end-to-end analytics support. With the advent of these tools, non-analysts can even be empowered to take control of low-level analytical tasks. Recent studies have shown that analytics-as-a-service technologies are advantageous for novices performing supervised tasks, but in their current form, can by no means replace the knowledge of a seasoned AP [25].

While automation may provide the "how" behind analytics, it still cannot provide the "why." Connecting math with real-world project context is a uniquely human skill, which takes several forms:

> **Knowing what questions to ask.** Identifying a business problem and translating it into an analytics task requires understanding of not only data, but also trends, questions, and concerns that may not be recorded in a spreadsheet. For example, in designing a recommender system you may want to minimize prediction error or optimize around revenue. The selection depends on the business goal.

> **Selecting the right method.** In cases where the best method for a task is unclear, experience and industry knowledge become paramount.

Interpreting and communicating model results.
Generating an insight, prediction, or recommendation
from data is only half of the battle; framing that information in an engaging way requires perception of the goals
of the recipient.

Should APs be concerned about the threat of automation running them out of a job? While automation is an incredible tool to
help APs be faster and more thorough in their work, it will not
replace APs any time soon. As for their abilities, automated platforms falter in two critical ways: connecting the real-world problem
to the data and utilizing experiential application knowledge.

7.3   SUMMING IT UP

Analytics

In this chapter, we begin by covering essential aspects of data. Then
we introduce how analytics is used to solve different challenges in
the world. The types of analytics are then presented. After this,
we review the analytics process. We rely heavily on the INFORMS
Body of Knowledge and use it as a reference for this section. The
next section covers selecting an analytical approach followed by
requirements gathering and model building. Model building is ex-
panded to cover different aspects of that process, including data
collection cleaning, data formatting and storage, and the development of the model components. Model testing and tuning are
covered next. To close the chapter, we provide guidance on the
deployment and management of models.

Trust

The trust chapter of the book discusses building trust through
defining roles in future collaborations, then focuses on different
types of trust including long-term and swift. We discuss trust in
the math model, a condition which is slightly different from trust-
ing the AP. We tie these two concepts together by highlighting that
the trust in the AP may be projected into the trust in the model because the audience may not understand the math and/or the data.
The following section covers interpretability and explainability. We
define interpretability as another AP being able to understand the
model and data. Explainability is defined as having a non-analytic

person understand the modeling data. The last topic we cover in this chapter is analytical ethics issues. We explain the difference between nefarious and unintentional errors and prescribe a method of documentation to ensure the AP is transparent and can avoid any perception of intentional unethical behavior.

Communication

In chapter 4, we begin with the importance of setting the context with any form of communication. Next, we move on to expectation management and consider what our target audience requires from our communication. We then cover the concept of explanations and how important it is to not only explain the "how" of the analytical solution but also explain the "why": the significance of the analytical solution to the real-world problem. Motivating the target audience to also see the "why" behind an analysis is critical for long-term success. We address the challenges of potential issues with math anxiety and ways to deal with this condition by creating atomized bits of knowledge. These smaller pieces of knowledge can then be exposed over time to provide background, context, and increase the trust factor in the math and the data. The next section covers some common mistakes that occur when APs are communicating technical work to technical audiences. We then summarize the levels of understanding and emphasize the importance of crafting your communication product to increase this level but not to overextend. The next section covers guiding the conversation, and we end the chapter with a short summary of data visualization as a tool for analytic communication.

Experience

In this chapter, we begin by introducing the concept of a specialist vs a generalist with respect to analytics. We then summarize the types of experience, which at the most general level include analytic experience and subject matter experience. Next, we cover learning through failure by reflecting on personal experience and by researching other historic analytical projects. The next section introduces ideas on how to gain experience when you're starting with none. We summarize a selection of key skills that cannot be automated in the final section.

Convince Them

Chapter 6 covers the final element of our equation: how all of the elements work together to achieve an end goal. In this chapter, we explore the pairwise relationships between our left-hand variables. Each analytics project will require a different mix of the elements, which may change over the course of the project. We explore definitions of success for an analytics solution from initial implementation to long-term use and cultural change. We then provide a fictional story to show one repetition and possible techniques applied to an analytical project.

7.4 PUTTING IT INTO PRACTICE

Now it is time for you to start applying actions that you have discovered in this book to your analytics professional life. No project is too small or too simple to start with in your analytics journey. Perhaps you are wondering what areas of your city have the most traffic to help minimize your commute time, or perhaps your spouse works as a nurse and their unit needs help with a shift scheduling tool. Opportunities to apply the lessons in this book are numerous if you observe the world around you with a fresh perspective.

Bibliography

[1] Google AI Blog. "Facets: An Open Source Visualization Tool for Machine Learning Training Data." Accessed June 22, 2022. http://ai.googleblog.com/2017/07/facets-open-source-visualization-tool.html.

[2] "OPTN: Organ Procurement and Transplantation Network - OPTN." Accessed June 22, 2022. https://optn.transplant.hrsa.gov/.

[3] Andrew White. "Our Top Data and Analytics Predicts for 2019," January 3, 2019. https://blogs.gartner.com/andrew_white/2019/01/03/our-top-data-and-analytics-predicts-for-2019/.

[4] Pedestrian Safety Action Plan. Technical report, City of Pittsburgh, Pittsburgh, PA, 2020.

[5] Laura Albert. I ended my talk about science communication with my answer to the questions: So what? Why do science communication & public outreach matter? @WomeninOR #WORAN #ORMS https://t.co/1luhin0jHs, January 2022.

[6] John C. Anderson and Thomas R. Hoffmann. A Perspective on the Implementation of Management Science. *The Academy of Management Review*, 3(3):563–571, 1978. Publisher: Academy of Management.

[7] Ross Anderson, Itai Ashlagi, David Gamarnik, Michael Rees, Alvin E. Roth, Tayfun Sönmez, and M. Utku Ünver. Kidney Exchange and the Alliance for Paired Donation: Operations Research Changes the Way Kidneys Are Transplanted. *INFORMS Journal on Applied Analytics*, 45(1):26–42, February 2015. Publisher: INFORMS.

[8] Linda Argote, Sunkee Lee, and Jisoo Park. Organizational Learning Processes and Outcomes: Major Findings and Future Research Directions. *Management Science*, 67(9):5399–5429, September 2021. Publisher: INFORMS.

[9] Linda Argote and Ella Miron-Spektor. Organizational Learning: From Experience to Knowledge. *Organization Science*, 22(5):1123–1137, September 2011.

[10] Joy Buolamwini and Timnit Gebru. Gender Shades: Intersectional Accuracy Disparities in Commercial Gender Classification. In *Proceedings of the 1st Conference on Fairness, Accountability and Transparency*, pages 77–91. PMLR, January 2018. ISSN: 2640-3498.

[11] Edited by James J. Cochran. *INFORMS analytics body of knowledge*. Wiley series in operations research and management science. Hoboken, NJ : John Wiley and Sons, Inc., 2019., 2019.

[12] Walter DeGrange and Wilson L. Price. Chapter 12: Why Won't They Use Our Model? In Natalie M. Scala and James P. Howard II, editors, *Handbook of Military and Defense Operations Research*, pages 268–280. Chapman and Hall/CRC, S.l., 1 edition edition, February 2020.

[13] Zandra Nymand Ennis, Dorthe Dideriksen, Henrik Bjarke Vaegter, Gitte Handberg, and Anton Pottegård. Acetaminophen for Chronic Pain: A Systematic Review on Efficacy. *Basic and Clinical Pharmacology and Toxicology*, 118(3):184–189, March 2016.

[14] Wai Fong Boh, Sandra A. Slaughter, and J. Alberto Espinosa. Learning from Experience in Software Development: A Multilevel Analysis. *Management Science*, 53(8):1315–1331, August 2007. Publisher: INFORMS.

[15] Steven L. Franconeri, Lace M. Padilla, Priti Shah, Jeffrey M. Zacks, and Jessica Hullman. The Science of Visual Data Communication: What Works. *Psychological Science in the Public Interest*, December 2021. Publisher: SAGE PublicationsSage CA: Los Angeles, CA.

[16] Jeremy Ginsberg, Matthew H. Mohebbi, Rajan S. Patel, Lynnette Brammer, Mark S. Smolinski, and Larry Brilliant. Detecting influenza epidemics using search engine

query data. *Nature*, 457(7232):1012–1014, February 2009. Bandiera_abtest: a Cg_type: Nature Research Journals Number: 7232 Primary_atype: Research Publisher: Nature Publishing Group.

[17] Dwayne D. Gremler and Kevin P. Gwinner. Rapport-Building Behaviors Used by Retail Employees. *Journal of Retailing*, 84(3):308–324, September 2008.

[18] Blake Hallinan and Ted Striphas. Recommended for you: The Netflix Prize and the production of algorithmic culture. *New Media & Society*, 18(1):117–137, January 2016. Publisher: SAGE Publications.

[19] Katherine Hawley. *Trust: A Very Short Introduction*. Oxford University Press, Oxford, 1st edition, September 2012.

[20] Constance L. Hays. What Wal-Mart Knows About Customers' Habits. *The New York Times*, November 2004.

[21] Mikael Holmqvist. Experiential Learning Processes of Exploitation and Exploration Within and Between Organizations: An Empirical Study of Product Development. *Organization Science*, 15(1):70–81, February 2004.

[22] Frens Kroeger, Girts Racko, and Brendan Burchell. How to create trust quickly: a comparative empirical investigation of the bases of swift trust. *Cambridge Journal of Economics*, 45(1):129–150, January 2021.

[23] Steve Krug. *Don't Make Me Think: A Common Sense Approach to Web Usability*. New Riders, Berkeley, Calif., revised edition edition, January 2014.

[24] E. S. Levine, Jessica Tisch, Anthony Tasso, and Michael Joy. The New York City Police Department's Domain Awareness System. *INFORMS Journal on Applied Analytics*, 47(1):70–84, February 2017. Publisher: INFORMS.

[25] Jasmien Lismont, Tine Van Calster, María Óskarsdóttir, Seppe vanden Broucke, Bart Baesens, Wilfried Lemahieu, and Jan Vanthienen. Closing the Gap Between Experts and Novices Using Analytics-as-a-Service: An Experimental Study. *Business & Information Systems Engineering*, 61(6):679–693, December 2019.

[26] Albert Fox Cahn and Will Luckman. Microsoft needs to stop selling surveillance to the NYPD, July 2020.

[27] Michael A. Martin. "It's Like... You Know": The Use of Analogies and Heuristics in Teaching Introductory Statistical Methods. *Journal of Statistics Education*, 11(2):1, January 2003.

[28] Emmanuel Martinez and Lauren Kirchner. The Secret Bias Hidden in Mortgage-Approval Algorithms – The Markup. Section: Denied.

[29] Leila Miller. LAPD will end controversial program that aimed to predict where crimes would occur, April 2020. Section: California.

[30] Koen Peters, Sérgio Silva, Rui Gonçalves, Mirjana Kavelj, Hein Fleuren, Dick den Hertog, Ozlem Ergun, and Mallory Freeman. The Nutritious Supply Chain: Optimizing Humanitarian Food Assistance. *INFORMS Journal on Optimization*, 3(2):200–226, January 2021.

[31] Frank C. Richardson and Richard M. Suinn. The Mathematics Anxiety Rating Scale: Psychometric data. *Journal of Counseling Psychology*, 19(6):551–554, 1972. Place: US Publisher: American Psychological Association.

[32] Harrison Schramm. Last Word: Taking Control. *Phalanx*, 54(3):56–61, 2021.

[33] Miriam W. Schustack and John R. Anderson. Effects of analogy to prior knowledge on memory for new information. *Journal of Verbal Learning and Verbal Behavior*, 18(5):565–583, October 1979.

[34] Thomas D. Vreugdenburg, Deanne Forel, Nicholas Marlow, Guy J. Maddern, John Quinn, Richard Lander, and Stephen Tobin. Morbidity and mortality meetings: gold, silver or bronze? *ANZ Journal of Surgery*, 88(10):966–974, 2018. _eprint: https://onlinelibrary.wiley.com/doi/pdf/10.1111/ans.14380.

Index